Liquid Fuels
from Coal

ACADEMIC PRESS RAPID MANUSCRIPT REPRODUCTION

Liquid Fuels
from Coal

Edited by

Rex T. Ellington

Fluor Engineers and Constructors, Inc.
Houston, Texas

ACADEMIC PRESS   New York   San Francisco London   1977
A Subsidiary of Harcourt Brace Jovanovich,Publishers

CHEMISTRY

16124-2111

ACADEMIC PRESS, INC.
111 Fifth Avenue, New York, New York 10003

United Kingdom Edition published by
ACADEMIC PRESS, INC. (LONDON) LTD.
24/28 Oval Road, London NW1

Library of Congress Cataloging in Publication Data
Main entry under title:

Liquid fuels from coal.

   Proceedings of a symposium held at the American
Chemical Society meeting in San Francisco, Aug. 29-
Sept. 3, 1976.
   Includes index.
   1. Coal liquefaction—Congresses. I. Elling-
tion, R. T. II. American Chemical Society.
TP343.L678   662'.6622   77-5002
ISBN 0-12-237250-6

PRINTED IN THE UNITED STATES OF AMERICA

CONTENTS

Contributors

Number in parentheses indicate page on which authors' contribution begin.

WILLIAM T. ABEL (19), U. S. Energy Research and Development Administration, Morgantown Energy Research Center, Morgantown, West Virginia 26505

SAYEED AKHTAR (183, 201), U. S. Energy Research and Development Administration, Pittsburgh Energy Research Center, 4800 Forbes Avenue, Pittsburgh, Pennsylvani 15213

R. L. BAIN (117), Colorado School of Mines, Chemical and Petroleum Refining Engineering, Golden, Colorado 80401

R. M. BALDWIN (177), Colorado School of Mines, Chemical and Petroleum Refining Engineering, Golden, Colorado 80401

RAND F. BATCHELDER (103), U. S. Energy Research and Development Administration, Pittsburgh Energy Research Center, 4800 Forbes Avenue, Pittsburgh, Pennsylvania 15213

JOSEPH R. COMBERIATI (19), U. S. Energy Research and Development Administration, Morgantown Energy Reasearch Center, Morgantown, West Virginia 26505

JAMES A. CUSAMANO (79), Catalytica Associates, Incorporated, 2 Palo Alto Square, Palo Alto, California 94304

ALAN DAVIS (255), The Pennsylvania State University, Coal Research Section, 517 Deike Building, University Park, Pennsylvania 16802

J. E. DOOLEY (221), U. S. Energy Research and Development Administration, Bartlesville Energy Research Center, P. O. Box 1398, Bartlesville, Oklahoma 74003

REX T. ELLINGTON, Fluor Engineers and Constructors, Incorporated, 4620 North Braeswood, Houston, Texas 77096

YUAN C. FU (103), U. S. Energy Research and Development Administration, Pittsburgh Energy Research Center, 4800 Forbes Avenue, Pittsburgh, Pennsylvania 15213

J. H. GARY (117), Colorado School of Mines, Chemical and Petroleum Refining Engineering, Golden, Colorado 80401

J. O. GOLDEN (117), Colorado School of Mines, Chemical and Petroleum Refining Engineering, Golden, Colorado 80401

JAMES A. GUIN (45, 133, 245), Department of Chemical Enginering, Auburn University, Auburn, Alabama 36830

HENRY W. HAYNES, JR. (63), University of Mississippi, Department of Chemical Engineering, University, Mississippi 38677

JOHN P. HENLEY (45), Department of Chemical Engineering, Auburn University, Auburn, Alabama 36830

CHAO-SHENG HUANG (63), University of Mississippi, Department of Chemical Engineering, University, Mississippi 38677

EDWIN S. JOHANSON (89), Hydrocarbon Research, Incorporated, P. O. Box 1416, Trenton, New Jersey 08607

DONALD R. JOHNSON (245), Department of Chemical Engineering, Auburn University, Auburn, Alabama 36830

CECILIA C. KANG (1, 89), Hydrocarbon Research, Incorporated, P. O. Box 1416, Trenton, New Jersey 08607

CLARENCE KARR, JR. (19), U. S. Energy Research and Development Administration, Morgantown Energy Research Center, Morgantown, West Virginia 26505

S. KATZ (173), Oak Ridge National Laboratorty, P. O. Box X, Oak Ridge, Tennessee 37830

R. C. KOLTZ (117), Colorado School of Mines, Chemical and Petroleum Refining Engineering, Golden, Colorado 80401

RICARDO B. LEVY (79), Catalytica Associates, Incorporated, 2 Palo Alto Square, Palo Alto, California 94304

GARETH D. MITCHELL (255), The Pennsylvania State University, Coal Research Section, 517 Deike Building, University Park, Pennsylvania 16802

T. O. MITCHELL (153), Mobil Research and Development Corporation, P. O. Box 1025, Princeton, New Jersey 08540

W. C. NEELY (245), Department of Chemical Engineering, Auburn University, Auburn, Alabama 36830

JOHN O. H. NEWMAN (183), U. S. Energy Research and Development Administration, Pittsburgh Energy Research Center, 4800 Forbes Avenue, Pittsburgh, Pennsylvania 14213

GOVANON NONGBRI (1), Hydrocarbon Research, Incorporated, P. O. Box 1416, Trenton New Jersey 08607

WALLACE S. PITTS (45, 133), Department of Chemical Engineering, Auburn University, Auburn, Alabama 36830

JOHN W. PRATHER (45, 133, 245), Department of Chemical Engineering, Auburn University, Auburn, Alabama 36830

B. R. RODGERS (173), Oak Ridge National Laboratory, P. O. Box X, Oak Ridge, Tennessee 37830

I. SCHWAGER (233), University of Southern California, Chemical Engineering Department, University Park, Los Angeles, California 90007

WILLIAM SPACKMAN (255), The Pennsylvania State University, Coal Research Section, 517 Deike Building, University Park, Pennsylvania 16802

NORMAN STEWART (1), Electric Power Research Institute, 3412 Hillview Avenue, Palo Alto, California 94304

GARY A. STYLES (45), Department of Chemical Engineering, Auburn University, Auburn, Alabama 36830

ARTHUR R. TARRER (45,133,245), Department of Chemical Engineering, Auburn University, Auburn, Alabama 36830

C. J. THOMPSON (221), U. S. Energy Research and Development Administration, Bartlesville Energy Research Center, P. O. Box 1398, Bartlesville, Oklahoma 74003

KUO-CHAO WANG (63), University of Mississippi, Department of Chemical Engineering, University, Mississippi 38677

MURRAY WEINTRAUB (201), U. S. Energy Research and Development Administration, Pittsburgh Energy Research Center, 4800 Forbes Avenue, Pittsburgh, Pennsylvania 15213

MILTON J. WEISS (201), U. S. Energy Research and Development Administration, Pittsburgh Energy Research Center, 4800 Forbes Avenue, Pittsburgh, Pennsylvania 15213

D. D. WHITEHURST (153), Mobil Research and Development Corporation, P. O. Box 1025, Princeton, New Jersey 08540

PAUL M. YAVORSKY (183, 201), U. S. Energy Research and Development Administration, Pittsburgh Energy Research Center, 4800 Forbes Avenue, Pittsburgh, Pennsylvania 15213

T. F. YEN (233), University of Southern California, Chemical Engineering Department, University Park, Los Angeles, California 90007

Preface

This book represents the efforts of seventeen authors to bring to the general and technical public new information on a broad range of topics on coal liquefaction. When the papers were first presented at the 172nd National Meeting of the American Chemical Society, there were many requests for joint publication. As a result, the authors put forth the extra effort to assemble this book. These papers and other recent contributions represent a healthy situation for technical progress beyond 1976 and real progress for commercial plant construction in the next decade if those opposing all energy development can be enlisted to help solve, rather than continue to deepen, the national energy predicament.

In placing the results of their work before their peers, these authors help bring into eclipse the situation that has existed for much of the last generation regarding growth of coal liquefaction technical information beyond that of World War II, or earlier, vintage. The intervening period can be characterized as one of a low level of development, which occurred in private industrial projects, Government facilities, and universities, in decreasing order. Only limited publication has been made of the fundamental or the commercial results of work during this period. Even more limited publication has been made of basic interpretations of reactions and kinetic models for commercial plant design. A number of the papers in this book present results and/or extensions of results known for some time to people inside the various earlier projects. Contrary to some criticisms regarding recent expenditures to "reinvent the wheel," they often represent valid independent confirmations, improvements, extensions, and certainly a willingness to stand before their peers for judgment of the quality of their product. These papers represent contributions to the following major areas of coal liquefaction knowledge:

1. basic dissolution and liquefaction processes,
2. kinetics of certain reactions,
3. important secondary processes, and
4. results of improved analytical methods.

Each contribution is the subject of brief comment to assist readers with limited years of study of the field in interrelating various results and conclusions.

The Editor's comments are made primarily from the aspect of

designing and constructing plants to produce coal liquids. Assessment of academic contribution or sophistication of approach will be left to others. Groups involved in the design of the commercial plants are in a difficult position. Pilot plant and demonstration plant runs are unlikely to have covered all the conditions that must be considered for optimized plant design. Unless a process owner/licensor can supply the necessary design data and is willing to guarantee process performance, the engineering/construction firm will be very reluctant to guarantee that the final plant will perform to meet the design conditions. Development and publication of technical data for wide ranges of conditions is a sound way to soften these problems.

People charged with commercial plant design must make many selections in balancing flow sheets, optimizing energy efficiency, and minimizing the cost of entire projects. Basic decisions depend on the desired product slate, required depth of conversion of the coal feed, and the number of process steps to be employed. One of the papers[1] presented at the same ACS symposium as the others in this book was unavailable for printing because of prior publication. This paper addressed some of the basic flow sheet balancing in commercial plant design. In this case, it suggests objectives for solvents and products for the primary liquefaction unit. The thesis of Gleim is that heptane-soluable product is the most valuable because it would eliminate the downstream refining problems caused by the presence of asphaltenes. The detrimental nature of asphaltenes is concluded from the problems they cause with coking and catalyst deactivation in the upgrading of residual petroleum fuels. Rough material balances were developed to show that only 50 to 60% of the coal need be converted to heptane-soluble product to balance a plant. This product is separable by distillation. The remainder of the coal is required for energy and hydrogen production. The lesson suggested for the process developer is that efforts to push processes to the point of liquefying 80% or more of the coal fed may be well beyond the point of diminishing returns. The greatest benefit may come from optimizing primary products at substantially lower overall liquefaction.

Gleim's basic ideas merit pondering by process developers and plant designers in a number of ways. For example, the advisability of putting more coal through the reactor to obtain the desired amount of product must be balanced against deeper conversion of a lesser amount. This decision can be made soundly only by enough work to develop mechanical flow sheets, specify and price equip-

[1]Gleim, W. K. T. "Liquefaction of Coal without Catalyst Using Selective Hydrogenation."

ment, and estimate final installed cost. The work should be done early to preclude some very bad decisions.

Whether one considers coal as a polymer or a condensed ring structure, the temperatures that must be employed to obtain effective solution rates cause the thermal rupturing of linkages that creates the primary products of greatest interest. Much has been accomplished in determining what affects production of the primary products but much remains to be done to describe the interrelationships involved. Secondly, similar data are required for the conversion or upgrading of primary products. Until such progress is made, including rates of reactions, industry may be foreclosed on true design optimization for the product slate desired by a particular client.

Certainly, with the comparisons available between low-temperature pyrolysis and solution, it is apparent that hydrogen or some other small free radical must be available to stabilize primary fragments as they are formed or condensation toward coke will occur. An effort to describe important characteristics of the solvent medium is described by Kang, Nongbri, and Stewart. This work focuses on the the tetralin content of the solvent as a measure of its hydrogen donor capacity. Temperature effects observed suggest that when thermal cracking exceeds hydrogenation donor capacity, coke formation will become significant.

An often-asked question is whether the first stage of a two-stage conversion should be catalytic, if two stages are judged to be necessary. For a given slurry oil, one might consider an empty reactor, one with fresh catalyst, one with spent catalyst, or perhaps the limiting case of a reactor filled with catalyst support type material, which might be termed a "thermal" reactor. Such a first step would logically be followed by a catalytic reactor on straight primary product or demineralized primary product. The paper of Karr *et al.* treats two cases of the thermal reactor, i.e., one filled with alumina and the other with silica material. The primary product is treated over CoMo catalyst. Those concerned with reaction models should also find the paper of Karr *et al.* helpful in selecting primary, secondary, and tertiary products. It is proper to foresee a generalized model that would fit all of the reactor fillings mentioned above.

In the SRC process the solvent/slurry oil is fractionated out of the product oil and recycled without pretreatment. Since the hydrogen donor capacity of the solvent is very important in maximizing light product and minimizing coking, the characteristics of the straight-run solvent become extremely important. Tarrer *et al.* have examined the effect of the minerals in the coal on the hydrogenation and hydrodesulfurization of creosote oil containing primarily phenanthrene, naphthalene, acenaphthene, and anthracene. If the hydrogenation of these components by the hydrogen in the dissolver can be increased, donor capacity can be increased. The demonstration that certain coal minerals do promote hydrogenation shows why the recycle of coal minerals to the reactor has been considered in some cases. Commercially, handling difficulties and relative ease of pretreating the solvent may preclude such recycle.

Whatever the initial process of solubilization of the coal at commercially feasible reaction temperatures, catalytic processes cannot become effective until a reasonable concentration is generated of fragments that are small enough to be adsorbed on catalyst particles without steric hindrance. It is generally accepted that fragment molecular weights of less than 300 to 500 are necessary for such conditions to exist. Analyses of primary liquefaction products have shown the existence of large concentrations of a number of polycyclic aromatic compounds. Obviously, improved knowledge of the behavior of these compounds during catalysis would facilitate work on catalysts, reactors, and the process design of commercial plants for the upgrading of primary coal liquids. It is pertinent to solvent regeneration or donor capacity boosting. A basic approach to expansion of such knowledge is to employ a clean simple system consisting of one of the primary products, such as phenanthrene. The paper by Haynes *et al.* examines the hydrogenation of phenanthrene at severities that will cover most commercial upgrading conditions for one catalyst. Catalyst selection was not a part of the investigation.

As process design considerations involve more complex systems it is natural to examine the desirability of using catalysts to augment production of desired materials from primary products in the first and/or second stage reactor. Either location imposes a severe environment on the catalyst. As a consequence, properties outside activity come into play in catalyst selection. A review of basic criteria for selection of catalysts for coal systems is provided by Levy and Cusamano. The review confirms selections already in use and suggests others for test.

Moving beyond the fundamental considerations of catalyst selection, Kang and Johanson examine the behavior of the old standby CoMo catalyst under the conditions of the H–coal reactor. Deactivation with processing of well over two thousand pounds of coal per pound of catalyst is described. The factors considered are carbon deposition, sintering, metals deposition, and attrition. Information of this type is helpful in evaluation of different processing approaches for selected products.

A new variation in the use of catalytic first stage reactors is that employing synthesis gas rather than hydrogen as discussed in the paper of Fu and Batchelder. The basic objective is laudable. If synthesis gas can be used for liquefaction, shift conversion and certain parts of gas clean-up may be eliminiated from the process unit normally required to produce hydrogen. The cost savings may be significant. These authors show that with the same operating conditions, comparable amounts of hydrogen and synthesis gas are consumed and comparable amounts of oil produced. The basic catalyst was CoMo-alkali carbonate. Analysis of the product liquid is not detailed except in following the concentration of oil and asphaltene, one of the accepted primary products. Flow testing in a Synthoil unit is not reported yet.

Good reaction equations and rate data and kinetic models based on these are mandatory if commercial plant design is to be optimized and performance guaranteed. These tools are certainly required unless every plant is to be designed right on a set of pilot or demonstration plant data. Data for hydrodesulfurization rates for a Kentucky coal dissolved in anthracene oil are provided by Koltz *et al.* The rates varied with time and temperature.

Among the many rates that may become pertinent to process design is the rate at which hydrogen can be made available for reaction in coal solution reactions. To this end Guinn *et al.* determined the rate at which hydrogen could be absorbed by a Kentucky coal-creosote oil system. The results indicate that the rate limiting step is the chemical consumption of dissolved hydrogen, not the process of solution.

Part of a significant study of rates is reported by Whitehust and Mitchell. It may overshadow many earlier investigations. Very short contact times and relatively complete product analyses are used to delineate fast early reactions and products. The results merit careful study for industrial process design. These include the rapidity of solution, early need for H-donor agent, delayed hydrogen consumption, and changes in product distribution. The reaction model with which the editor has had greatest comfort over the years is that related to Given. It suggests that depolymerization of the basic coal structure occurs upon solution yielding a molecular weight distribution peaking in the 5000 range. Thermal reactions then yield fragments with molecular weights in the 300 to 700 range. The Whitehurst and Mitchell work shows an early production of more than 2000 MW followed by rapid production of 300 to 900 MW material. They show rapid early yield of SRC-type product with little hydrogen consumption. With a push for deeper conversion, hydrogen consumption increases markedly.

One of the toughest engineering problems facing the industrial process designer is related to the mineral matter and undissolved coal in the reactor product. Somehow this material must be removed from the final product. Distillation is not cheap, even when applicable. Among the approaches being explored are particle agglomeration, filtration, high gravity and magnetic fields, and antisolvent methods. Katz and Rodgers have tried to reform the inorganic particulates to facilitate their separation. Solvent dilution with process-generated solvent improved filterability only as much as a 40ºC increase in temperature. Neither of these approaches or surface active agents yielded separability of the type needed in plants.

In another paper, Newman et al. explore use of a process-generated antisolvent to promote agglomeration of solid particles. With a short period of treatment filterability was improved substantially without need for filter-aid. Processes of this type may offer considerable promise.

Finally, Weintraub et al. tested a number of filtration approaches. None would yet seem to overcome the extreme operational and economic problems associated with commercial usuage of filtration.

The keystone of reaction modeling and good material balances is in the existence of good methods for analyzing different plant streams. Many product streams in coal liquefaction are extremely complex and in years past analyses would take longer than the experiment. In many cases good reaction equations could not be prepared because analytical methods were not available or were prohibitively expensive. Recently analytical methods have taken such a leap forward that the opportunity for delineation of coal liquefaction reactions is enhanced greatly.

An extremely detailed procedure is reported by Dooley and Thompson using many procedures developed for the analysis of high-boiling petroleum fractions. The analytical system is based on physical and chemical separation followed by instrumental characterization. From this and other recent advances process developers should be able to select procedures that will realize full value from their reaction investigations.

Schwager and Yen employed solvent fractionation to obtain oil, resin, asphaltene, carbene, and carboid cuts. Color index, solvent elution chromatography, X-ray, NMR, and silica gel chromotography were employed to analyze the cuts.

Prather et al. discuss high-pressure liquids chromatography. With the suite of methods now available, it should be possible to prove or disprove old reaction concepts and follow the lives of important consitituents in future work.

Finally, optical studies of residue particles from coal liquefaction are described by Mitchell, Davis, and Spockman. Petrographic examination of coals yields much information on source material and reactive parts thereof. Association of feed coal macerals with residue components may provide yet another way to trace changes coal during processing.

The second paper[2] ineligible for this book because of prior publication describes fast pyrolysis of small coal particles in an atmosphere of hydrogen. It is a logical extension of free fall hydrogenolysis of coal and oil shale done some years ago. A significant aspect of the work of Pelofsky *et al.* is that preliminary balances of commercial type flow sheets show that product yields of only about one-half the feed coal (lignite) are necessary to balance the plant flow sheet. Cracking severity can be varied to shift the proportions of liquid and gaseous product.

Most significantly, the liquid product is predominantly BTX with minor amounts of naphtalene, toluene, and anthracene, and the gaseous product is primarily methane and ethane. This product slate provides the potential for a relatively simple plant with only a few, high-priced products.

To conclude, coal technology is advancing in many areas. It is hoped that joint presentation of these papers will help show how the efforts fit together.

[2]Pelofsky, A. H., Greene, M. I., and LaDelfa, C. J., "Short Residence Time (SRT) Coal Hydropyrolys".

THE ROLE OF SOLVENT IN THE SOLVENT REFINED COAL PROCESS

Cecilia C. Kang and Govanon Nongbri
Hydrocarbon Research, Inc.
(A subsidiary of Dynalectron Corporation)

Norman Stewart
Electric Power Research Institute

A preliminary study of the effect of solvent upon SRC
process performance was undertaken. The startup solvent
and makeup solvent have a significant effect upon coal
conversion. However, the use of a good startup solvent
does not sustain good process performance under unfavorable process conditions and vice versa. Hydroaromatics, represented by H_β from NMR analysis, other
than tetralin also possess hydrogen donor capabilities.
These hydroaromatics could be better donors than tetralin. This conclusion was reached during a process
study of the Black Mesa coal, a subbituminous coal. At
the same hydrogen pressure, an increase in reactor temperature resulted in higher hydrogen consumption, lower
coal conversion and higher dry gas (C_1-C_3) production.
These observations were accompanied by a deterioration
in solvent properties, (as shown by the lower tetralin
to naphthalene ratio and lower H_β value) and significant amount of coke formation. These findings support
a hypothesis that coke formation results when thermal
cracking gets ahead of hydrogenation which is catalyzed
by the hydrogen donors present in the solvent.

I. INTRODUCTION

The coal-derived liquid used to produce a pumpable liquid
slurry feed to hydrogenation systems is commonly referred to
as slurry oil or pasting oil. The role of this material,
other than as a transport medium, has been neglected until the
last few years. The pasting oil used in Bergius Catalytic
Hydrogenation consisted of fractions collected at various
points in the downstream process. Some of these products of

carbonization and/or distillation were catalytically hydrogenated before being used. No specific analyses were made
which provided information about a chemical basis for understanding hydrogen donor capability. Necessary solvent qualities that contribute to or enhance coal conversion, hydrogenation or desulfurization were not quantified. Pott-Broche
used a hydroaromatic middle oil generated from coal or tar
hydrogenation as their pasting oil and were able to achieve
conversions of about 80% at pressures in the 1500-3000 psig
range.

More recently, processes that depend upon the hydrogen
donor capacity of the process solvent have been developed.
Among these are the Consol Synthetic Fuel Process (CSF), Pamco
Solvent Refined Coal Process (SRC), and Exxon Donor Solvent
(EDS) Process.

The CSF and EDS processes depend upon special solvent
production through fractionation and subsequent hydrogenation.
The SRC process does not employ solvent preparation other than
fractionation.

Interest has grown in the role of solvent in these three
processes as bench scale and small Pilot Plant investigations
have intensified. The application of analytical techniques
toward identifying donor hydrogen was described by R.P.
Anderson (1). More recently, donor reactions for desulfurization were described by G. Doyle (2).

The Exxon (EDS) process in operation since 1975 at the
one-ton-a-day scale depends heavily upon knowledge of solvent
quality and the ability to control it (3). A proprietary solvent quality index was reported by Exxon. A minimum quality
index reported to be a function of liquefaction conditions was
related to conversion and claimed to improve handling qualities
of the products.

This work reports some rather dramatic bench scale coal
processing data. Major effects are attributed to identifiable
slurry solvent properties.

II. THE ROLE OF SOLVENT

In conjunction with a process study undertaken at HRI
under Electric Power Research Institute Research Project 389
to investigate SRC process operability and product yield
structures for several coals of commercial interest, a preliminary analysis of the role of solvent upon coal conversion
was carried out. The purpose of this project was to screen
coals prior to their testing at the 6 T/D SRC Pilot Plant at
Wilsonville, Alabama. The operation of this plant has been
jointly funded by EPRI and Southern Services, Inc. Catalytic
Inc. is the plant operator. The study was undertaken with two

subbituminous coals, Wyodak coal and Black Mesa coal, and one
bituminous coal, Illinois No. 6 coal from the Monterey Mine.
The proximate and ultimate analyses of these three coals are
summarized in Table 1. All of the work described here was

TABLE 1
Analysis of Feed Coals
==

	Wyodak Coal	Black Mesa Coal	Illinois No. 6 Monterey Coal
Proximate Analysis (dry basis) W %			
Ash	7.04	10.10	10.11
Volatile Matter	46.48	42.57	40.85
Fixed Carbon	46.48	47.33	49.04
Ultimate Analysis (dry basis) W %			
Carbon	67.78	69.81	69.71
Hydrogen	4.97	4.79	4.56
Sulfur	0.80	0.33	4.52
Nitrogen	0.65	1.04	1.17
Ash	7.04	10.10	10.11
Oxygen (by difference)	18.76	13.93	9.89
Sulfur Forms (dry basis) W %			
Pyritic Sulfur	0.15	0.17	1.23
Sulfate Sulfur	0.01	0.00	0.14
Organic Sulfur (by difference)	0.51	0.20	2.78
Total Sulfur	0.67	0.37	4.15
Mineral Analysis (ignited basis) W %			
P_2O_5	0.35	0.13	0.34
SiO_2	27.91	46.99	50.25
Fe_2O_3	5.30	5.33	18.66
Al_2O_3	15.75	16.96	18.15
TiO_2	1.10	0.96	0.87
CaO	19.00	18.00	4.25
MgO	5.60	2.24	0.86
SO_3	22.84	6.17	3.29
K_2O	0.48	0.81	1.92
Na_2O	0.72	1.69	1.17
Undetermined	0.95	0.72	0.24

done in a continuous bench-scale non-catalytic unit with a

reactor having a volume of 1000 cc. This unit has been described previously in EPRI reports 123-1-0, 123-2, and 389-1
(4,5,6,7), and is shown in Fig. 1.

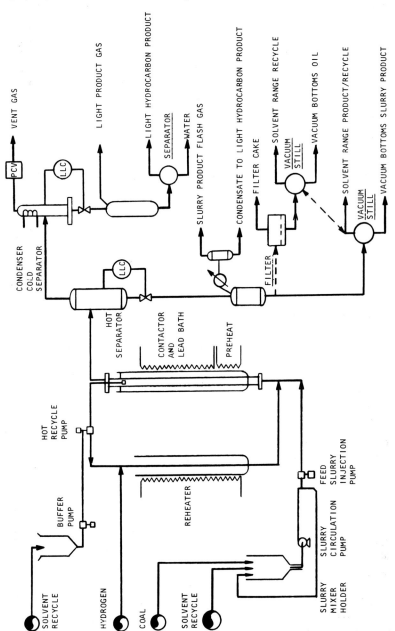

Fig. 1. Bench Scale Unit Flow Diagram

The solvent was characterized by its tetralin to naphthalene ratio as determined by gas chromatographic analysis, and the H_β value by NMR measurement. H_β represents hydrogen on carbon atom at β-position of the aromatic ring (excluding methyls).

A. Effect of Tetralin Content of Solvent upon Coal Conversion Wyodak Coal Study

During the SRC processing of Wyodak coal it was observed that, under steady operating conditions, there was a gradual but persistent change in coal conversion with the operating time. A plot is presented in Fig. 2 to show process conditions and coal conversion versus days of operation. This experimental program was carried out under four conditions as shown. The startup solvent and recycle solvent were characterized by tetralin content and tetralin to naphthalene ratio. Unconverted coal is defined as the benzene insoluble organic component of dry filter cake. This value is obtained by filtering the slurry product in the bench scale filter at 300°F,

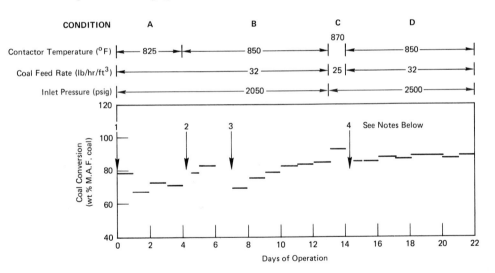

Fig. 2. The effect of tetralin content of solvent upon coal conversion.

Notes: 1. Startup solvent contained 2.8% tetralin, and 0.24 tetralin to naphthalene ratio. 2. About 3000 to 5000 grams of the startup solvent were added during the period of interruption. 3. Wyodak coal operations solvent generated during preceding periods, containing about 1% tetralin, was used for startup. 4. Startup solvent was hydrogenated anthracene oil, containing 1.3% tetralin and 0.76 tetralin to naphthalene ratio.

and extracting the cake with benzene to quantify the benzene insoluble content. The dry filter cake value is corrected by subtracting the benzene insolubles present in the 975°F+ oil. Coal conversion in this paper is derived from unconverted coal by using a forced ash balance for bituminous coal or a SO_3-free ash balance for subbituminous coal.

Under Condition A, the startup solvent contained 2.8% tetralin and a tetralin to naphthalene ratio of 0.24. Period 1 gave 79% coal conversion which dropped to 68% during Period 2 as tetralin content decreased to 1.1%. These changes indicate that tetralin content in a startup solvent may not by itself sustain good coal conversion under unfavorable process conditions.

Condition B was at higher contactor temperature than Condition A. The effect of the tetralin content of the start-up solvent was more fully demonstrated under Condition B; Periods 5 and 6 yielded higher conversion than Period 8. These two intervals were at the same operating conditions, but were interrupted by a shutdown caused by unit plugging. Period 8 was started with a solvent which was generated in Condition A, containing only 1% tetralin, but Period 5 was preceded with an addition of 3,000 to 5,000 grams of the startup solvent on Condition A containing 2.8 tetralin to the unit loop with a total holdup of 11,000 grams.

The gradual, but persistent, increase in coal conversion observed during Periods 8 through 13 was rather unique in that it had not been observed before. This was probably caused by the use of a poor startup solvent produced from preceding periods at lower reactor temperatures, which contained low tetralin. The solvent from Period 13 had a tetralin content of 1.3%. The bench unit has a rather large holdup capacity; at 15% solvent production, it takes about six days to displace 70% of the startup solvent. These observations further indicate that the hydrogen donor content of the solvent as well as coal conversion is controlled by the process conditions.

Condition D used hydrogenated anthracene oil as the start-up solvent. Raw anthracene oil, a coal tar product, was hydrogenated to increase its hydrogen content from 5.5 to 7.2% in a separate operation prior to its use in this work. The hydrogenated anthracene oil has a higher initial boiling point (426°F) than all the other solvents used during this study. Hence, its tetralin (boiling point - 405°F) content is only 1.3%. However, its exceedingly high tetralin to naphthalene ratio, 0.76, indicates that it would be a good donor solvent if it contained other high boiling hydroaromatics with donor capacity. Condition D was at higher pressure than Condition B, 2500 psig vs 2050 psig. Hence, Condition D was expected to yield higher coal conversion than Condition B. The use of

hydrogenated anthracene oil as startup solvent did result in a
high initial conversion of   The coal conversion only increased slightly to 88% in contrast to the steady increase
from 70% to 85% under Condition B which used a poor startup
solvent possessing a tetralin to naphthalene ratio of about
0.16. The high initial conversion and slight increase in conversion observed under Condition D leads to a conclusion that
the hydrogenated anthracene oil is a better donor solvent than
the other startup solvent and recycle solvent because it contains high concentrations of hydrogen donors other than tetralin.

B. The Effect of Other Hydrogen Donors - Black Mesa Coal Operation

The presence of hydrogen donors other than tetralin in
the coal-derived solvents is confirmed during the study of
Black Mesa coal.

Solvent analyses for the Black Mesa operations are summarized in Table 2. A plot is presented in Fig. 3 to show
process conditions and coal conversion.

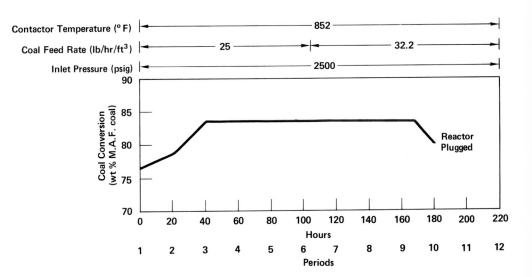

Fig. 3. The effect of other hydrogen donors upon coal
conversion as shown by the Black Mesa coal study.

TABLE 2
SRC Operations on Black Mesa Solvent Analysis
===

Run Number 177-	Startup Solvent From Wilsonville	Recycle Solvent			
		119-1	-2B	-5	-7/8
Reactor Temperature, °F (estimated)		847	845	838	841
Inlet Pressure, psig		2500	2500	2500	2500
Coal Feed Rate, Lbs/Hr/Ft^3		21.4	25	25	32.2
Coal Conversion, W % of M.A.F. Coal		77.1	79.2	83.0	83.3
Hydrogen Consumption W % of M.A.F. Coal				2.94	2.84
Simulated Distillation (VPC) Boiling Point, °F					
IBP, °F	284	302	293	302	266
IBP-403°F, W %	10.3	13.3	15.3	11.7	13.2
405°F, (Tetralin)	4.2	5.1	1.6	2.5	2.5
412°F (Naphthalene)	12.1	11.2	13.2	10.9	11.5
413-642°F	50.2	50.0	47.8	52.3	48.7
644°F (Phenanthrene/ Anthracene)	9.9	7.2	7.3	6.6	8.2
646°F-EP	13.4	13.2	14.8	16.0	15.9
EP, °F	797	806	833	833	824
Tetralin+Naphthalene	16.3	16.3	14.8	13.4	14.0
Tetralin/Naphthalene	0.35	0.46	0.12	0.23	0.22
Proton Distribution (H-NMR) [a]					
H_{ar} (9-6 ppm)	52.9	53.5	54.2	50.3	48.2
H_{α} (3.5-2 ppm)	24.2	24.3	24.2	23.7	23.9
H_{β} (2-1.1 ppm)	17.5	16.5	16.5	20.5	21.6
H_{γ} (1.1-0.4 ppm)	5.4	5.7	5.1	5.5	6.3

a. H_{ar} represents aromatic hydrogens, H_{α} benzylic hydrogens including methyls, H_{β} hydrogens on carbon atoms at β-positions of aromatic rings, and H_{γ} hydrogens in aliphatic methyl groups.

The Black Mesa coal operation showed a steady increase in coal conversion which then remained at this level during the rest of the run. This increase in coal conversion was out of phase with the tetralin content and the tetralin to

naphthalene ratio of the solvents. The startup solvent was
obtained from the Wilsonville Unit using another coal. It
contained 4.2% tetralin and tetralin to naphthalene ratio of
0.35. These values were much higher than those of the recycle
solvent which contained 1.6 to 2.5% tetralin and a tetralin to
naphthalene ratio of 0.12 to 0.23. However, the coal conversion steadily increased from 77 to 83.5% and   remained
at this level. An examination of the NMR data disclosed that
H_β increased steadily from 16.5 to 21.6% from Period 1 to
Period 7/8. Since the tetralin content of the startup solvent
was higher than that of the recycle solvent, the increase in
H_β represents an increase in other hydrogen donors through the
displacement of the startup solvent. These observations confirm that the solvent produced from the hydrogenation of Black
Mesa coal contains other hydrogen donors, which may be more
reactive towards Black Mesa coal than tetralin.

C. The Effect of Solvent Quality upon Coke Formation –
    Monterey Coal Operation

    Wilsonville started operation on Illinois No. 6 Monterey
coal in August 1975. Since the Monterey coal had a higher
organic sulfur content (2.8%) than the Illinois No. 6 coal
from Burning Star Mine which had been used previously, the
dissolver temperature was raised from the previous 835°F level
to 855-890°F at 1700 psig in an attempt to produce SRC meeting
the sulfur specification of 0.96 weight percent. There was an
indication of solids buildup in the dissolver during these runs.
This was confirmed by flushing out 400-500 pounds of solids
from the dissolver, which has a total volume of 20 cubic feet.
Later, extensive plugging of transfer lines around the dissolver occurred with substantial coke accumulation in the lower
half of the dissolver. The plugging forced a shutdown of the
plant. HRI was retained by EPRI to carry out bench unit SRC
experiments under conditions similar to Wilsonvilleь. Experiments were carried out at 1500 psig, 840 and 868°F contactor
temperatures. The coking problem was demonstrated and dupli-
cated in the bench-unit operation at the higher temperatures.
    Table 3 summarizes solvent analyses together with certain
pertinent operating data. At contactor temperature of 840°F,
the recycle solvents contained more hydroaromatics than the
startup solvent from the Wilsonville SRC plant, as shown by the
higher tetralin/naphthalene ratio and higher NMR H_β measurements of the recycle solvents. Hence, the recycle solvents
from this run should be better solvents than the Wilsonville
solvent. This assumption was supported by the slight, but
steady, increase in coal conversion during this run.

TABLE 3
SRC Operations on Illinois No. 6 Monterey Mine Coal
Solvent Analysis
===

Run Number 177	Startup Solvent From Wilsonville	Recycle Solvent	
		117-4	118-3B
Contactor Temperature, °F (Estimated)		839	863
Inlet Pressure, psig		1500	1500
Coal Feed Rate, Lbs/Hr/Ft^3		25	25
Coal Conversion, W % of M.A.F. Coal		94.0	89.5
Hydrogen Consumption, W % of M.A.F. Coal		4.1	5.1
Simulated Distillation (VPC) Boiling Point, °F			
IBP, °F	284	320	320
IBP-403°F, W %	10.3	14.8	15.0
405°F (Tetralin)	4.2	5.2	4.1
412°F (Naphthalene)	12.1	10.0	11.3
413-642°F	50.2	49.3	46.0
644°F (Phenanthrene/ Anthracene)	9.9	5.3	6.3
646°F End-Point	13.4	15.4	17.3
End Point, °F	797	824	860
Tetralin + Naphthalene	16.3	15.2	15.4
Tetralin/Naphthalene	0.35	0.52	0.36
Proton Distribution (H-NMR)[a]			
H_{ar} (0-6 ppm)	52.9	48.0	54.6
$H\alpha$ (3.5 - 2 ppm)	24.2	27.0	24.5
$H\beta$ (2 - 1.1 ppm)	17.5	18.3	16.4
$H\gamma$ (1.1 - 0.4 ppm)	5.4	6.7	4.5

a. H_{ar} represents aromatic hydrogen, $H\alpha$, benzylic hydrogen including methyls, $H\beta$, hydrogen on carbon atoms once removed from aromatic rings (excluding methyls), and $H\gamma$, hydrogen in aliphatic methyl groups.

Plots are presented in Fig. 4 to show coal conversion, hydrogen consumption, sulfur content of 350°F+ fuel oil, and dry gas production.

MONTEREY COAL

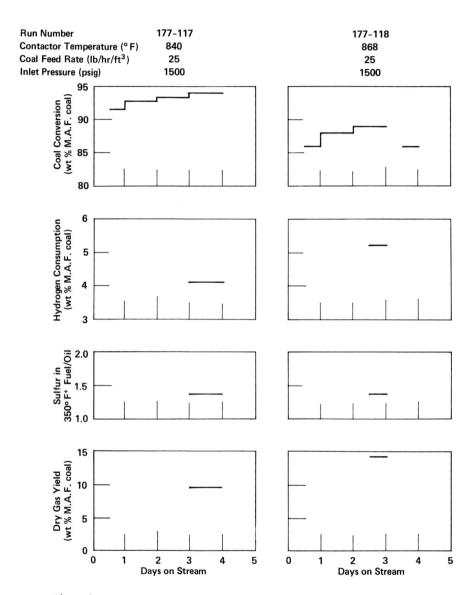

Run Number	177-117	177-118
Contactor Temperature (°F)	840	868
Coal Feed Rate (lb/hr/ft³)	25	25
Inlet Pressure (psig)	1500	1500

Fig. 4. The effect of process temperature upon product distribution.

The hydrogen consumption was higher at high contactor temperature, 5.1 vs. 4.1 W %, as expected. However, the high hydrogen consumption was accompanied by unfavorable product distribution as follows:

1.   Low coal conversion, 89.5% at high temperature vs 94% at low temperature.

2.   High C_1-C_3 dry gas production, 13.3% vs 7.8%.

3.   A deterioration in solvent properties, as shown by the low tetralin to naphthalene ratio, 0.36 vs 0.5, and low $H\beta$ Value, 16 vs 19%.

4.   High Coke Formation: The run at 840°F was operated fcr 83 hours without reactor plugging. The run at 868°F was operated for 78 hours then terminated by reactor plugging. The reactor was filled with a hard plug of coke between the middle of preheater to a length of 9 feet out of reactor length of 13 feet.

These phenomena lead to the hypothesis as follows: Coal hydrogenation is promoted by the hydrogen donors present in the solvent. When the process conditions result in a deterioration of the donor capability of the solvent, thermal cracking gets ahead of hydrogenation, which can lead to coke formation.

D.   Coke Formation

1.   Characterization of Coke

Two coke samples taken from the Wilsonville dissolver after operations on Illinois No. 6 coal from Monterey Mine (8) were sent to Pennsylvania State University for optical characterization. One was a fine-grained material and the other was a much coarser grain size. Penn State reported (9) that the fine-grained sample was composed of insoluble particles surrounded by a layer of anisotropic carbon. The most common constituents of the nuclei are semi-coke and calcite, both of which are present in feed coal. The shell of anisotropic carbon is deposited on these nuclei in the SRC process with this coal. Fine-grain anisotropy and extinction characteristics are indicative of a uniform, onion-like deposition, i.e., concentric layers surrounding a nucleus. The coarse grain sample exhibited agglomeration of the fine-grained deposit and there was a coarsening of anistropy of the carbonaceous shell. These domains or mosaic structures are recognized as the more traditional optical texture of coalesced mesophase.

The immediate cause of sedimentation of the particles in the reactor was the growth of anisotropic carbon on to the undissolved materials. These insolubles consist of a semi-coke contaminant, mineral matter and undissolved macerals (generally

fusinitic material). The resulting adhesion of particles
forms a coke-like sediment. Operating conditions and optical
texture of the bonding anisotropic carbon suggest that its
growth from the solution is by a mechanism of growth of nematic
liquid crystals, thus resulting in mesophase formation which
then leads to a non-plastic semi-coke.

Coke analyses from both Wilsonville and HRI operations
with Illinois No. 6 Monterey coal are summarized in Table 4.

Most of the coke recovered from the dissolver and various
parts of the Wilsonville unit (such as the high pressure let-
down valves and the blowdown tank) were similar in composition.
These compositions are represented by a UCC (undissolved carbonaceous cresol insolubles) to ash ratio of around 2 with the
exception that the coke which deposited at the vapor-liquid
interface contained much more cresol-insoluble carbonaceous
material than the others. The coke recovered from the HRI
bench units had a much higher ash content than those from the
Wilsonville unit. The higher ash content of coke, together
with the shorter time of HRI operation, coincides with Penn
State's characterization of coke as a uniform, onion-like
carbon deposition surrounding a nucleus of mineral matter.

The H/C atomic ratios of all coke deposits (including the
coke deposits at the vapor-liquid interface) were quite consistent, about 0.50. This is significantly lower than the H/C
atomic ratio of the Monterey Mine coal, 0.78, and the unconverted coal, 0.73. There was no difference in the H/C atomic
ratio between the coke accumulated over short periods of time
(three to four days in the HRI unit) and long periods of time
(more than one month in the Wilsonville unit). The consistency
of the H/C of coke under these circumstances leads to the pos-
tulation of a reprecipitation mechanism. After the coke deposition, it did not undergo any significant carbonization.

2. Mechanism for Coke Formation

The SRC process as practiced at Wilsonville uses a recycle
solvent (350°F to 750°F) for transporting the coal into the
reactor and for promoting the dissolution and the conversion
of coal. The effect of solvent upon coke formation is illustrated by the following reactions. Reaction (1) illustrates
the ability of a donor solvent molecule to release donor hydrogen during liquefaction using tetralin as an example of a donor
solvent molecule.

$$\text{(DONOR SOLVENT)} \xrightarrow{\text{Liquefaction}} \text{(SPENT SOLVENT)} + 4H \qquad (1)$$

DONOR SOLVENT   SPENT SOLVENT

TABLE 4
Analysis of Coke Deposits
==

Coal Feed	Wilsonville Operation				
	Monterey				
Date Sampled	8/28/75	8/28/75	8/31/75	9/4/75	9/5/75
Sample No.	9464	9343	---	9440	9457
Run No.	--	--	--	--	--
Location	Let Down Valve LV-415 550-600°F	Low Pressure Flash at Pump P-136	Flushing-Out Solids Blow-Down Tank BD-144	Top of Dissolver Vapor-Liquid Interface	Bottom of Dissolver 6' cone
Ash, W %	31	34	35	19	33
UCC (1), W %	64	58	63	75	66
Cresol Soluble, W %	5	8	2	7	1
UCC/Ash	2.06	1.71	1.80	3.89	2.00
Benzene Insoluble, W %	--	--	--	--	--
Benzene Soluble, W %	--	--	--	--	--
Benzene Insoluble/Ash	--	--	--	--	--
Sulfur, W %					
Sulfide	1.68	1.41	1.80	0.79	0.55
Pyritic	0.14	0.10	0.18	0.13	0.33
Sulfate	0.11	0.50	0.12	0.16	0.09
Organic	0.64	1.52	0.57	0.62	1.34
Elemental Analysis, W %					
C	60.83	56.12	56.95	73.55	
H	2.53	2.14	2.27	3.02	
N	1.03	1.09	1.02	1.15	
Cl	0.78	0.46	0.96	1.36	
O (by difference)	0.79	2.06	1.28	0.84	
S Total	2.57	3.54	2.67	1.70	2.50
Ash	31.47	34.63	34.86	19.28	
H/C Atomic Ratio	0.50	0.46	0.48	0.49	

(1) Undissolved carbonaceous cresol insolubles

TABLE 4
Analysis of Coke Deposits
===

Coal Feed	Wilsonville Operation Burning Star		HRI Operation Monterey		
Date Sampled	4/7/75	4/8/75	--	--	
Sample No.	--	--	--	--	
Run No.	--	--	177-117	177-118	
Location	Dissolver Solids Flushout	Dissolver Solids Hydroblasted	Dissolver	Dissolver Middle Section	Dissolver 2' to Bottom Section
Ash, W %			52.26	28.99	41.37
UCC (1), W %					
Cresol Soluble, W %					
UCC/Ash					
Benzene Insoluble, W %			19.18	44.75	44.92
Benzene Soluble, W %			28.56	26.26	13.71
Benzene Insoluble/Ash			0.37	1.54	1.09
Sulfur, W %					
Sulfide					
Pyritic					
Sulfate					
Organic					
			Benzene-Insoluble Solids		
Elemental Analysis, W %					
C	69.09	42.99	34.40	55.26	47.50
H	2.02	1.86	1.35	2.19	1.78
N	1.00	0.73			
Cl	0.10	0.05			
O (by difference)	2.95	0.81	--	--	--
S Total	0.87	5.24	5.61	3.55	3.58
Ash	24.07	48.62	52.26	28.99	41.37
H/C Atomic Ratio	0.51	0.52	0.47	0.48	0.45

(1) Undissolved carbonaceous cresol insolubles

The coal undergoes chemical transformation in the temperature range of 570 to 930°F giving rise to light boiling materials and a solid residue. Reaction (2) shows the formation of low boiling materials and free radicals of heavy coal fragments from thermal cracking of the coal "molecules" and the acceptance of donor hydrogen by the free radicals to form liquid product.

$$\text{coal} \xrightarrow{\text{heat}} \begin{array}{c}\text{low boiling}\\\text{materials}\end{array} + \begin{array}{c}(A_r\text{-X})\\\text{coal fragment}\end{array} \xrightarrow{\text{donor H}} \begin{array}{c}A_r\text{-XH}\\\text{liquid product}\end{array} \qquad (2)$$

X = C, O, S, or N

When there is a deficiency of donor hydrogen the coal fragments recombine to form coke as indicated by reaction (3).

$$n(A_r\text{-X}) \longrightarrow \begin{array}{c}(A_r\text{-X})n\\\text{coke}\end{array} \qquad (3)$$

The presence of solid particles, such as mineral matter, acts as a nucleus upon which the reprecipitation of coal fragments occurs.

III. SUMMARY AND CONCLUSIONS

A preliminary study of the effect of solvent upon SRC process performance was undertaken. As shown by the Wyodak Coal Study, the startup solvent and makeup solvent have a significant effect upon coal conversion. However, the use of a good startup solvent does not sustain good process performance under unfavorable process conditions and vice versa. Hydroaromatics were measured in part by H_β from NMR analysis. Hydroaromatics other than tetralin also possess hydrogen donor capabilities. These hydroaromatics could be better donors than tetralin. These conclusions were reached during a process study of the Black Mesa coal which is a subbituminous coal. A Monterey coal process study demonstrated that at the same hydrogen pressure, an increase in reactor temperature resulted in higher hydrogen consumption, lower coal conversion and higher dry gas (C_1 - C_3) production. The poor process performances were accompanied by a significant amount of coke formation and a deterioration in solvent properties (as shown by the lower tetralin to naphthalene ratio and lower H_β value). These findings support a hypothesis that coke formation results when thermal cracking gets ahead of hydrogenation which is promoted by the hydrogen donors present in the solvent.

IV.   ACKNOWLEDGEMENTS

The work described here was done as part of Electric Power Research Institute Research Project RP -389. The purpose of the work was to generate process operability., yield structure and hydrogen consumption data for candidate feed coals to the Utility Industry Solvent Refined Coal Test Unit at Wilsonville, Alabama.

The stimulating discussions and solvent analysis provided by Dr. D. D. Whitehurst and Dr. T. O. Mitchell (whose work was carried out as part of EPRI Research Project 410) are sincerely appreciated.

Analysis of coke from the Wilsonville unit was carried out under EPRI Research Project 366 at the Pennsylvania State University.

V.   REFERENCES

1.   Anderson, R.P., "Evolution of Steady State Process Solvent in the Pittsburgh & Midway Solvent Refined Coal Process", Symposium on Coal Processing, American Institute of Chemical Engineers, Salt Lake City, Utah, August 20, 1974.

2.   Doyle, Gerald, "Desulfurization via Hydrogen Donor Reactions", Progress in Processing Synthetic Crudes and Resids, American Chemical Society Chicago Meeting, August 24-29, 1975.

3.   Furlong, L.E., Effron, E., Vernon, L.W., Wilson, E.L., "Coal Liquefaction by the Exxon Donor Solvent Process", The 1975 National AIChE Meeting, Los Angeles, November 18, 1975.

4.   "Solvent Refining of Coal", Electric Power Research Institute Report, RP-123-1-0, August 1974.

5.   "Liquefaction of Kapairowits Coal", Electric Power Research Institute Report, RP123-2, October 1974.

6.   "Solvent Refining of Illinois No. 6 and Pittsburgh No. 8 Coals", Electric Power Research Institute Report, EPRI 389, Vol. I, June 1975.

7.   "Solvent Refining of Wyoming, Illinois No. 6 (Monterey Mine) and Black Mesa Coals", Electric Power Research Institute Report, EPRI 389, Vol. II, May 1976.

8.   "Solvent Refined Coal Pilot Plant, Wilsonville, Alabama", SRC Technical Report No. 7, January 1976.

9.   "Characterization of Mineral Matter in Coals and Coal Liquefaction Residues", Pennsylvania State University Report, RP366-1, December 1975.

PRODUCTS FROM TWO-STEP COAL LIQUEFACTION
USING THREE DIFFERENT FIRST-STEP REACTOR PACKINGS

Clarence Karr, Jr., William T. Abel, and Joseph R. Comberiati
U.S. Energy Research and Development Administration
Morgantown Energy Research Center

The results of a series of runs, with a laboratory-scale stirred batch reactor system, and a laboratory-scale continuous flow reactor system, designed to approximate the behaviour of a two-step coal hydroliquefaction process, are presented. The results of detailed chemical and physical characterization of the products obtained with a slurry of coal and catalytically hydrogenated tar oil (representing recycle oil), using either vitrified ceramic, alpha-alumina, or silica for a nominally non-catalytic first-step reactor packing, are presented. The results of detailed characterization of the products obtained by catalytic hydrogenation of the filtered first-step product in a second-step reactor are presented, and the effects of the different second-step reactor feeds on the activity of the catalyst are described. Probable mechanisms for the functions of the first-step packing are described, as well as differences in the activity of the cobalt molybdate catalyst.

I. INTRODUCTION

In the current work, laboratory-scale batch and flow studies have been conducted with various nominally non-catalytic materials as reactor packing for the first-step of a two-step coal hydroliquefaction process being studied at the Morgantown Energy Research Center's Liquid Fuels Research and Development Branch. An examination of the available literature on some three dozen coal liquefaction processes gave no evidence for any previous studies along these lines. Several of the processes have been compared in review articles (1,2). The process studied is a two-step coal/recycle oil slurry feed hydroliquefaction process, with removal of solid residue between the two steps, in which the second-step catalytic reactor is preceded by a first-step reactor containing a nominally non-catalytic material selected for its ability to induce desireable hydropyrolytic reactions. These reactions are principally the conversion of asphaltene intermediates from the coal feed to oils and/or to asphaltenes of higher hydrogen content. Unconverted asphaltenes will likely result in coke formation in the catalytic reactor, greatly reducing catalyst activity and selectivity. Hydropyrolytic treatment (thermally induced dissociation or disproportionation of hydrogen) of petroleum residues, before a catalytic desulfurization, is designed to break down the less thermally stable asphaltic compounds so as to give less coke formation on the desulfurization catalyst (3). The presence of a high surface area carbonaceous material apparently accelerates the desired reactions, as in the "Combifining" process in which asphaltic compounds are treated with hydrogen at 30 atm in a fluidized-bed of petroleum coke at 380-410°C (4). The deactivation effect of coal asphaltenes is reduced by hydropyrolytic splitting on the surface of semi-coke dust (5). Pore diameters between 100 and 1,000 angstroms, and larger, appear necessary to allow free access of asphaltenes for their conversion. Conversely, when only small diameter pores are present, asphaltenes are reported to block the pores and the material is rapidly deactivated (6).

The material desired as first-step reactor packing for the process studies must, therefore, in consideration of the pertinent literature, have a moderate surface area with a surface structure allowing or promoting carbon deposition, high porosity, and large pores (0.1 micron, or larger). A nominally noncatalytic material is indicated, to restrict surface reactions to those induced by a carbonaceous deposit. This would appear to rule out silica-alumina and possibly gamma-alumina, while allowing alpha-alumina and silica. Vitrified ceramics would apparently not offer sufficient surface area or porosity, although they are nominally non-catalytic for this reason.

Dissolution of the coal in the first-step reactors allows separation of the mineral residues. Although coal minerals have been suspected to have substantial coal-liquefaction catalytic behavior (7), these same minerals, or trace metals in them, act as strong poisons for hydrogenation catalysts (8). In addition, the sub-micron, clay-derived mineral residues plug the pores of the catalyst. Massive plugging between catalyst pellets occurs when the mineral residues settle out. In the initial developmental work on the Synthoil Process for making low-sulfur fuel oil from coal it was demonstrated that this type of plugging could be prevented by a highly turbulent flow of hydrogen through the catalyst bed (9). However, if the density of the product oil is reduced from that of a fuel oil to that of a kerosine and gasoline rich product, massive

plugging occurs even with highly turbulent hydrogen flow, because the light kerosine and gasoline flash off into the gas
phase, leaving insufficient liquid vehicle to carry residual
solids through the fixed bed.

By introducting an effective hydropyrolytic step followed
by separation of the mineral matter, the catalyst, such as the
widely used cobalt molybdate catalysts on silica-promoted
alumina supports, can have a longer life, and can be used at
the milder temperatures and pressures for which they were designed, rather than at more rigorous conditions necessary for
trying to maintain some degree of activity.

With regard to pressure, as the hydrogen pressure increases, there is a maximum in the overall yield of hydrocracking (10). In a two-step coal hydrocarbonization-tar vapor catalytic hydrogenation laboratory-scale study, the maximum yield of product was obtained at a hydrogen pressure of
40 atm (11). With an increase in temperature above 400°C,
thermal splitting by a free radical mechanism starts to be
significant, leading to the formation of coke-like materials.
High hydrogen pressures will reduce coking to some extent, but
are economically unfavorable, while resulting in decreased
yield of liquid product. At 375°-400°C and 1,500 psig, catalytic hydrogenation of a low-temperature coal tar gave a product with high aliphatic character and a high diesel index
(12).

II. EXPERIMENTAL

A. Stirred Batch Reactor Runs

Coal-oil slurries were made up with one part by weight of
run of mine coal (Pittsburgh No. 8 seam, Ireland mine, 33.06
weight percent volatile matter, 0.72 weight percent moisture,
19.51 weight percent high-temperature ash, carbon 60.42, hydrogen 4.32, nitrogen 0.92, oxygen 7.96, and sulfur 4.62
weight percent, all by analysis) and two parts of hydrogenated
Reilly tar oil. This hydrogenated oil was prepared in a one-
gallon stirred batch reactor with hydrogen gas at 390°C and
1,800 psig for three hours, using about 1,800 ml oil and about
32 g presulfided cobalt molybdate on silica-promoted alumina
1/8-inch pellets (Harshaw CoMo-0402T 1/8") in two baskets attached to the stirrer, as shown in Figure 1. The catalyst was
presulfided in situ with a flow of 10-15 percent hydrogen sulfide in hydrogen at 3-4 liters per hour per 100 g catalyst for
1.5 hours at 400°C and atmospheric pressure. Gas chromato-
graphic/mass spectrometric analysis showed about 20 percent

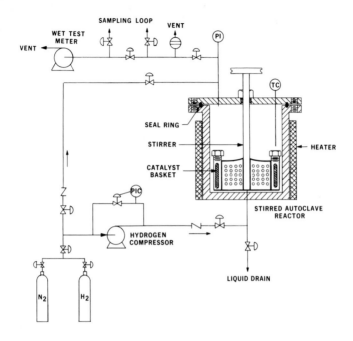

Fig. 1.   Stirred autoclave reactor system.

identifiable hydroaromatics, or hydrogen donors, in the hydrogenated oil, compared to none in the original oil. The identities of these are discussed under Section III.

These coal/oil slurries were examined for their behavior
over two different kinds of packing that could be tried in the
first-step reactor of the two-step process. These were a
vitrified ceramic (Norton "Denstone 57" catalyst bed support,[1]
consisting of 1/4-inch balls having a silica content of 56.4
weight percent, with a surface area of about 0.01 m^2/g and a
very low porosity), and alpha-alumina (Girdler catalyst carrier T-375, consisting of 1/8-inch pellets having a silica
content of only 0.02 weight percent, with a surface area of
about 5.3 m^2/g, and 0.06-0.8 micron pores). A series of runs
was made in a one-gallon stirred batch reactor, using 36.15 g
of "Denstone" or alpha-alumina in two baskets attached to the
stirrer, and about 900 g slurry at 450°C and 1,800 psig for
three hours, or about 1,800 g slurry at 430°C and 1,500 psig
for three hours, the reactor being brought up to the desired
pressure with hydrogen gas. Air was flushed from the system
with nitrogen gas and nitrogen purged from the system with

---

[1]   Use of trade names or company names is for identification
only and does not imply endorsement by ERDA.

hydrogen gas before partial pressurizing, heating, and then
any final pressurizing. As hydrogen was consumed the pressure
was maintained with additional hydrogen.

Half of each first-step product from all runs was filtered to remove mineral residue, and the filtered and unfiltered
portions were then hydrogenated over the same catalyst used to
prepare the hydrogenated tar oil (Harshaw CoMo-0402T 1/8").
The second-step run conditions for all eight runs were identical, namely, 1,500 psig (achieved with hydrogen gas) at 380°C
for one hour in a one-gallon stirred reactor, using about 200
g feed and 0.4 g presulfided cobalt molybdate on silica-promoted alumina in two baskets attached to the stirrer. The
quantity of catalyst was chosen to approximate 500 hours operation at a liquid hourly space velocity of one in a fixed bed
process.

The products were analyzed by 1) solvent extraction to
recover benzene insolubles, asphaltenes (benzene-soluble,
cyclohexane-insoluble), and oils (cyclohexane-soluble); 2)
liquid elution chromatography of the oils from activated alumina with cyclohexane and benzene (to remove colored resins);
3) gas chromatographic/mass spectrometric analysis of the
cleaned oil to identify and quantify individual compounds; 4)
elemental analysis for determination of the atomic hydrogen-
to-carbon ratio in various samples; 5) gas chromatographic
analysis of the gaseous products for the amounts of the individual hydrocarbons; and 6) physical property determinations
(density and distillation curves) on the filtered products.
Differences due to different operators, procedures, or equipment were avoided. The separation procedures were similar to
those previously used for analysis of low-temperature coal
tars (13).

B. 0.2 Pounds Coal/Hour Flow System

A schematic flow diagram of this system is detailed in
Figure 2. Interchangeable reactors packed for either first-
step runs, or with catalyst for second-step runs, were used in
the clam shell heaters. This laboratory-scale equipment was
used in a run of nearly 48 hours duration to prepare about
five gallons of hydrogenated tar oil for slurry preparation.
The Reilly tar oil was fed at a rate of about 375 ml/hr (LHSV
1.4) through presulfided cobalt molybdate on silica-promoted
alumina at 380°C and 1,500 psig hydrogen. Product sample collected at 44 hours analyzed for 26.26 weight percent hydroaromatics, while the hydrogen donor content for the composite
five gallon product was 26.88 weight percent, showing no significant decline in catalyst activity over about 48 hours
operation. The tar oil feed contained no hydrogen donors.

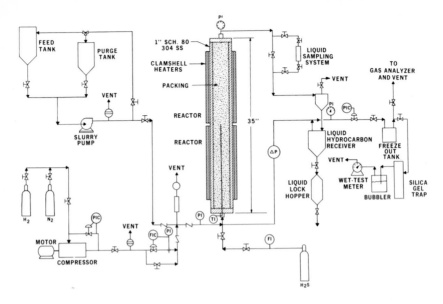

Fig. 2.   Laboratory-scale flow system.

Hydropyrolytic first-step runs were made in the same reactor (1-inch, schedule 80, 304 stainless steel pipe), with
the first seven inches packed with 1/2-inch lengths of 1/4-
inch O.D. 304 stainless steel tubing and the remaining 21
inches packed with either 1/8-inch alpha-alumina pellets, or a
99.5 weight percent alpha-quartz packing (Girdler catalyst
carrier T-1571, consisting of 3/16-inch silica spheres with a
surface area of about 131 m^2/g, and 1-18 micron pores).   The
total volume of the reactor was 270 ml.   The slurry feed was
made from one part by weight of 70 percent minus 200 mesh, 100
percent minus 100 mesh, Ireland Mine coal (Pittsburgh No. 8
seam), and three parts of the hydrogenated Reilly tar oil,
made as described above.   The slurry was fed at a rate between
245 and 375 g/hr at 440°C, 1,500 psig, and a hydrogen flow
rate of about 7.0 scfh.   Total run time with each first-step
packing was nearly 16 hours.   The faster feed rate for the bed
packed with spheres compensated for the larger void volume for
the spheres.

The products from both runs were filtered to remove mineral residue, and the residue washed with benzene to remove
adhering oil, dried, and weighed.   The filtered products were
then subjected to catalytic hydrogenation in the second-step
reactor under conditions essentially the same as those used to
prepare the hydrogenated tar oil.   The second-step reactor was
identical to the first-step reactor, except for the packing,

TABLE 1
Mass Balance for Stirred Batch Reactor Runs

| Run conditions | Grams In | | | | Grams Out | | | Unmonitored component[c] | | | | Ratio total grams out/ total grams in |
	Coal	Hydrogenated oil or 1st stage product	H₂ gas	Total	Liquid product	Tars[a]	Monitored gases[b]	NH₃	H₂S	H₂O	Total	
First-step[d] Denstone extreme	298.70	597.30	24.24	920.20	792.10	5.45	82.08	0.19	15.69	20.89	916.40	0.9959
First-step[e] Denstone mild	601.30	1202.70	26.45	1830.45	1673.69	16.90	53.27	-1.12	20.51	35.83	1798.99	0.9828
First-step[d] alpha-alumina extreme	307.67	612.33	19.41	939.41	791.70	24.50	78.41	-0.12	12.80	21.79	929.08	0.9890
First-step[e] alpha-alumina mild	599.20	1198.60	13.22	1811.02	1635.90	23.15	64.31	-3.14	21.83	36.89	1778.94	0.9823
Second-step[f] Denstone extreme	---	233.30	13.23	246.53	222.60	---	19.19	0.11	0.06	-0.04	241.92	0.9813
Second-step Denstone mild	---	233.00	13.23	246.23	209.50	---	15.98	0.22	0.43	0.79	226.92	0.9216
Second-step alpha-alumina extreme	---	222.00	13.22	235.22	198.20	---	12.70	1.00	0.22	0.45	212.57	0.9037
Second-step alpha-alumina mild	---	213.00	13.22	226.22	205.40	---	14.88	0.41	0.32	-0.08	220.93	0.9766

a. Residual products sticking to equipment.
b. Monitored gases: H_2, CO, CH_4, C_2H_6, CO_2, C_2H_4, C_3H_8, C_4H_{10} by gas chromatography.
c. Unmonitored components: NH_3, H_2S, H_2O. Amounts of these components were calculated from the differences between input and output N, S, and O.
d. Hydropyrolytic first-step with Denstone vitreous ceramic packing, at 1,800 psig and 450°C.
e. 1,500 psig, 430°C.
f. All four second-step runs with the cobalt molybdate catalyst, at 1,500 psig and 380°C, using unfiltered first-step product as feed.

25

TABLE 2
Hydrogen Balances for Stirred Batch Reactor Runs

Run conditions	Grams H_2 In				Grams H_2 Out					Ratio total grams out/ total grams in	Grams, H_2 gas in, corrected
	Coal	Hydrogenated oil or 1st stage product	H_2 gas	Total	Liquid product	Tars[a]	Monitored gases[b]	Unmonitored components[c]	Total		
First-step Denstone extreme[d]	12.90	41.81	24.25	78.96	55.45	0.22	18.98	3.27	77.92	0.9868	23.21
First-step Denstone mild[e]	25.98	84.19	26.45	136.62	112.13	0.68	12.90	5.19	130.90	0.9581	20.73
First-step alpha-alumina extreme[d]	13.29	42.86	19.41	75.56	54.23	1.47	19.31	3.17	78.18	1.0347	22.03
First-step alpha-alumina mild[e]	25.89	83.90	13.22	123.01	113.04	0.92	14.28	5.38	133.62	1.0863	23.44
Second-step Denstone extreme[f]	--	16.33	13.22	29.55	16.03	--	13.51	0.02	29.56	1.0003	13.23
Second-step Denstone mild	--	15.61	13.22	28.83	15.21	--	14.56	0.16	29.93	1.0382	14.32
Second-step alpha-alumina extreme	--	15.21	13.22	28.43	13.81	--	12.54	0.24	26.59	0.9353	11.38
Second-step alpha-alumina mild	--	14.72	13.22	27.94	14.17	--	14.64	0.09	28.90	1.0344	14.18

a. Residual products sticking to equipment.
b. Monitored gases: CH_4, C_2H_6, C_2H_4, C_2H_2, C_3H_8, C_4H_{10} by gas chromatography.
c. Unmonitored components: NH_3, H_2S, and H_2O. Amounts of H_2 consumed by forming these components were calculated from the difference between input and output N, S, and O.
d. Hydropyrolytic first-step with Denstone vitreous ceramic packing, at 1,800 psig and 450°C.
e. 1,500 psig, 430°C.
f. All four second-step runs with the cobalt molybdate catalyst at 1,500 psig and 380°C, using unfiltered first-step product as feed.

TABLE 3
Carbon Balances for Stirred Batch Reactor Runs

Run conditions	Grams C In			Grams C Out				Ratio total grams out/ total grams in
	Coal	Hydrogenated oil or 1st stage product	Total	Liquid product	Tars[a]	Monitored gases[b]	Total	
First-step[c] Denstone extreme	180.33	539.36	719.69	706.71	5.18	54.21	766.10	1.0645
First-step[d] Denstone mild	363.01	1086.04	1449.04	1496.65	16.06	36.23	1548.94	1.0689
First-step[c] alpha-alumina extreme	185.74	552.93	748.67	707.46	22.05	52.46	781.97	1.0445
First-step[d] alpha-alumina mild	361.74	1082.34	1441.18	1463.64	21.99	44.97	1530.60	1.0599
Second-step[e] Denstone extreme	---	208.15	208.15	200.36	---	5.61	205.97	0.9895
Second-step Denstone mild	---	208.12	208.12	187.50	---	1.04	188.54	0.9059
Second-step alpha-alumina extreme	---	198.38	198.38	179.39	---	0.13	179.52	0.9049
Second-step alpha-alumina mild	---	190.57	190.57	184.80	---	0.19	184.99	0.9707

a. Residual products sticking to equipment.
b. Monitored gases: CO, CH₄, C₂H₆, CO₂, C₂H₄, C₃H₈, C₃H₆, C₄H₁₀ by gas chromatography.
c. Hydropyrolytic first-step with Denstone vitreous ceramic packing, at 1,800 psig and 450°C.
d. 1,500 psig, 430°C.
e. All four second-step runs with the cobalt molybdate catalyst at 1,500 psig and 380°C, using un-filtered first-step product as feed.

TABLE 4

Solvent Extraction Analysis of Product from Two-Step Coal Liquefaction in Stirred Batch Reactor[a]

First-step run conditions	1,800 psig, 450°C, 3 hrs				1,500 psig, 430°C, 3 hrs			
	Not filtered between steps		Filtered		Not filtered between steps		Filtered	
Mineral residue First-step reactor material Fraction	Denstone	Alpha-alumina	Denstone	Alpha-alumina	Denstone	Alpha-alumina	Denstone	Alpha-alumina
Benzene insolubles	11.3	11.5	9.2	10.5	12.3	14.4	15.0	18.6
Asphaltenes (benzene-soluble, cyclohexane-insoluble)	9.3	5.2	26.1	11.6	16.3	14.4	26.0	16.5
Oils[c] (cyclohexane-soluble)	79.4	83.3	64.7	77.9	71.4	71.2	59.0	64.9

a. Figures in weight %.

b. Second-step run conditions of all eight runs: 1,500 psig, 380°C, 1 hr, presulfided cobalt-molybdenum on silica-promoted alumina.

c. Includes the bulk of the oil used to make up the coal/oil slurry (ratio 1:2).

which was entirely sulfided cobalt molybdate on silica-promoted alumina. The liquid feed was introduced at a rate between 250 and 350 g/hr at 380°C, 1,500 psig, and a hydrogen flow rate of about 7.0 scfh. Total run time was up to about 11 hours.

III. RESULTS AND DISCUSSION

A. Stirred Batch Reactor Runs

Mass, hydrogen, and carbon balances for the stirred batch reactor runs, including data on the four first-step runs and the four corresponding second-step runs with unfiltered first-step product as feed, are presented in Tables 1 through 3. The process conditions for all stirred batch reactor runs are summarized in Table 4.

Mass balances varied between 90.4 and 99.6 percent, with an average of 96.7 percent. Six of the eight runs had mass balances between 97.7 and 99.6 percent. Ammonia, hydrogen sulfide, and water were determined by calculating the decrease in elemental nitrogen, sulfur, and oxygen in going from solids and liquids in the feed to those in the product, and converting to the equivalent weights of the three compounds. Essentially all the water, and nearly all of the hydrogen sulfide, were evolved in the first-step runs. The alpha-alumina appeared to favor a greater production of water and hydrogen sulfide than the Denstone vitrified ceramic, except for hydrogen sulfide at the more rigorous conditions.

Hydrogen balances varied between 93.5 and 108.6 percent, with five of the eight runs between 98.7 and 103.8 percent. Hydrogen usage was estimated from the pressure gauges on the cylinder regulators, but temperature variations restricted the accuracy of this approach because the cylinders were mounted outside the high pressure cell building. A better estimate of hydrogen usage was obtained by assuming a balance of 100 percent and correcting the grams H_2 gas in, accordingly. The gas chromatographic analyses of product gases showed that the alpha-alumina favored a greater production of hydrocarbon gases than the Denstone vitrified ceramic. This coupled with the greater production of H_2O and H_2S would be expected to require a greater consumption of hydrogen with alpha-alumina. The corrected values for grams H_2 gas in come much closer to showing this than the uncorrected values, which appear to be generally unreliable.

Carbon balances varied between 90.5 and 106.9 percent, with half the values between 97.1 and 106 percent. The amounts of CO and CO_2 in the product gases from the first-step runs appeared to be somewhat high compared to the previous work, and may account for the 104.5 to 106.9 percent carbon recoveries for the first-step runs, as opposed to the 90.5 to 99 percent recoveries for the second-step runs.

The mass, hydrogen, and carbon balances for the stirred batch reactor runs fall within the range of values generally found for such laboratory-scale operations.

The results of the solvent extraction analysis of the products from the stirred batch reactor runs are summarized for all eight runs in Table 4, with the data for the paired "Denstone" and alpha-alumina first-step reactor materials placed together for ease of comparison. A large, but noncalculable, part of the oils comes directly from the compounds in the oil used to make the coal/oil slurry, while the rest of the oil comes from hydroliquefaction of the coal and the asphaltic constituents. Assuming that the lowest oil yield from the coal was only about one percent, with almost 90 percent recovery of vehicle oil (10 percent conversion to gas and light oils), it can be seen that in most instances that amount of oils was increased from 20 percent to as much as three-fold by using alpha-alumina instead of vitrified ceramic. In one instance, even though the oil yields were essentially the same the atomic hydrogen-to-carbon ratio for the alpha-alumina-derived oil was much higher, as shown later, demonstrating a higher total hydrogen gain for the alpha-alumina-derived oil.

In line with the larger amounts of coal-derived oil, using the alpha-alumina, larger amounts of identifiable coal-derived compounds were obtained (Table 5), with the compounds arranged according to increasing boiling point. These six compounds are all polycyclic aromatic hydrocarbons which were not detectable in the hydrogenated tar oil, and therefore probably came from the coal via the asphaltenes. The slightly lower concentration at the higher temperature and pressure may be explained by the dilution with a little more of other coal-derived compounds. There was about 50 or 60 percent more coal-derived, or more correctly, asphaltene-derived compounds in the second-step product oil using first-step product from the alpha-alumina bed as feed. Possibly this is because this feed did not deactivate the cobalt molybdate catalyst as much as the first-step product from the vitrified ceramic.

The higher activity of alpha-alumina for producing low-boiling hydrocarbons directly from coal or asphaltenes was also demonstrated by the gaseous hydrocarbons collected during the first-step runs. The volume of methane, ethane, propane, and butanes per pound of coal was in each instance greater for each hydrocarbon compound when using alpha-alumina (Table 6). The yields of hydrocarbon gas are also greater at 450°C than at 430°C, because this is the temperature range in which bituminous coals show a rapid increase in thermal decomposition.

TABLE 5

Coal-Derived Compounds and Alkylated Compounds in Two-Step Products,
Using Filtered Feed for Second-Step[a]

First-step run conditions	1,800 psig, 450°C		1,500 psig, 430°C	
First-step reactor material	Denstone	Alpha-alumina	Denstone	Alpha-alumina
Trimethylbenzenes	0.40	0.40	0.23	0.08
Methyl-, Dimethyl-, and Ethylnaphthalenes	11.58	14.04	9.95	9.73
Methylbiphenyls	0.43	0.44	0.32	0.37
Methyldibenzofurans	3.10	2.91	2.71	3.65
Methylfluorenes[b]	*1.07*	*1.16*	*0.85*	*1.40*
Methylphenanthrenes	2.60	2.92	2.79	2.93
Methylpyrenes	0.45	0.50	0.35	0.54
Benz(c)phenanthrene[b]	*0.10*	*0.14*	*0.18*	*0.37*
Methylchrysene[b]	*0.18*	*0.42*	*0.09*	*0.23*
Benz(b,j,k)fluoranthene[b]	*0.19*	*0.53*	*0.29*	*0.40*
Benz(a & e)pyrene[b]	*0.11*	*0.17*	*0.20*	*0.25*
Methylbenz(a & e)pyrene[b]	*0.04*	*0.07*	*0.04*	*0.06*
Total coal-derived compounds	1.69	2.49	1.65	2.71
Percent Increase[c]	47.3		64.3	
Total alkylated compounds	20.54	22.86	17.33	18.99
Percent Increase	11.3		9.6	

   a. Wt %, calculated on the basis of the oil in the coal/oil
slurry.
   b. Compounds not detected in the hydrogenated tar oil used
for slurry, hence derived from coal.
   c. Percent increase in going from Denstone to alpha-alumina
packing.

The greater yields of hydrocarbon gas with alpha-alumina are not due to increased dealkylation of alkylated polycyclics in the oil (Table 5). In fact, somewhat greater yields of alkylated compounds were obtained at both operating temperatures, using alpha-alumina.

TABLE 6
First-Step Hydrocarbon Gas Yields[a]
==

Run conditions	1,800 psig, 450°C		1,500 psig, 430°C	
Reactor material	Denstone	Alpha-alumina	Denstone	Alpha-alumina
Methane	1.645	1.890	0.835	0.985
Ethane	0.619	0.810	0.246	0.308
Propane	0.274	0.449	0.131	0.188
Butanes	0.043	0.070	0.022	0.025
Total	2.581	3.219	1.234	1.506
Percent Increase	24.7		22.2	

a. Figures in scf/lb coal.

The yields of asphaltenes (Table 4) were less with alpha-alumina, generally down to about one-half the quantity obtained with the vitrified ceramic. In the one instance in which the yields were close, the atomic hydrogen-to-carbon ratio was substantially higher for the alpha-alumina-derived asphaltene (Table 7). Under both of the first-step run conditions the atomic hydrogen-to-carbon ratios for the asphaltenes were clearly higher using alpha-alumina, as shown in Table 7. In one instance, the oil derived in the presence of alpha-alumina was distinctly higher in atomic H/C. In the other, the two oils had nearly identical values of atomic H/C. The yield of oil in the latter was greater with alpha-alumina (Table 4), demonstrating a higher total hydrogen uptake.

The total amounts of nine important classes of alkylated polycyclic aromatics were greater for the products obtained using alpha-alumina (Table 5), under both first-step run conditions. Larger amounts were obtained with both first-step materials at the more rigorous conditions, due to more reaction of the coal, and the percent increase for alpha-alumina was considerably greater.

TABLE 7
Elemental Analysis of Two-Step Products[a]

First-step runs	1,800 psig, 450°C		1,500 psig, 430°C	
First-step reactor material	Denstone	Alpha-alumina	Denstone	Alpha-alumina
Benzene insolubles	0.62	0.65	0.72	0.73
Asphaltenes	0.64	0.70	0.75	0.78
Oils	0.95	0.94	0.95	0.99

a. Atomic H/C (unfiltered feed for second step).

The identities and amounts of the eight identifiable hy-droaromatics, or hydrogen donors, in the hydrogenated tar oil used to make up the slurry for the stirred batch reactor runs are shown in Table 8. Analysis of the oil in the two differ-ent first-step products showed nearly double the consumption of hydrogen donors in the presence of alpha-alumina, compared to the vitrified ceramic.

It is possible that the much greater surface area and pore volume of the alpha-alumina compared to the vitrified ceramic could offer more carbonaceous surface for the hydrogen donors to react. Examination of the spent materials visually, and by scanning electron microscopy, showed that a black, car-bonaceous deposit covered the exterior of the ceramic balls, but extended throughout the interior of the alpha-alumina pellets.

In all instances, the presence of mineral residue in the second-step reactor gave improved results (Table 4) as regards decreased yields of asphaltenes and increased yields of oils. As mentioned in the Introduction, this apparent advantage is outweighed by several disadvantages.

Because the slurry feed to the stirred batch reactor runs was 67 percent by weight of hydrogenated tar oil, the distil-lation curves to a final boiling point of 495°C of the first-step products and the vehicle oils were rather similar. How-ever, some differences were evident which reflected the dif-ferences in activity of the two different reactor packings and the two different reactor conditions, as shown in Table 9. The largest cut, the median boiling diesel oil fraction, was essentially identical for the two tar oils and three of the four products. The fourth product, obtained with the alpha-alumina packing under the rigorous conditions, showed a sub-stantial decrease in this major fraction, with a corresponding substantial increase in the lowest molecular weight fraction, the gasoline cut. At the milder conditions, there was only a little more gasoline range fraction with the alpha-alumina packing than with the vitrified ceramic packing, and as ex-pected both packings gave substantially less gasoline cut than under the more rigorous conditions.

TABLE 8

Hydrogen Donors in Hydrogenated Tar Oil for Slurry Before and After Reaction with Coal: Analysis of First-Step Products[a]

| Type reactor | Batch 1,800 psig, 450°C | | | Flow 1,500 psig, 440°C | | |
| Reactor conditions | | | | | | |
Reactor packing Hydrogenated tar oil	Lot A	Denstone	Alpha-alumina	Lot B-F	Silica	Alpha-alumina
Compound						
Indan	1.22	1.51	1.25	1.63	0.72	0.83
Methylindans	0.35	2.94	2.59	0.40	0.32	0.60
Tetralin	7.34	6.61	4.62	9.11	6.58	6.37
Methyltetralins	1.93	2.09	1.00	4.29	3.87	3.35
Dihydrophenanthene	3.08	1.32	1.11	2.00	1.28	1.17
Octahydrophenanthrene	--	--	--	2.04	1.66	1.89
Tetrahydrophenanthrene	2.42	1.18	1.12	4.30	4.09	2.92
Tetrahydropyrene	3.27	0.59	0.84	2.11	0.44	0.45
Hexahydropyrene	--	--	--	0.20	0.12	0.12
Dihydropyrene	0.97	0.60	0.69	0.80	0.40	0.40
Total	20.58	16.84	13.22	26.88	19.48	18.10
Percent reacted		18.2	35.8		27.5	32.6

a. Figures in weight %.

TABLE 9
Distillation of Slurry Vehicle Oils Compared with Distillation of Filtered
First-Step Products from Stirred Batch Reactor Runs[a]

Reactor conditions	1,800 psig, 450°C			1,500 psig, 430°C		
Reactor packing			Alpha-			Alpha-
Hydrogenated tar oil	Lot A	Denstone	alumina	Lot B-F	Denstone	alumina
Distillate range, °C						
40-202 (gasoline)	10.9	17.1	21.0	13.8	11.1	12.2
202-265 (kerosine)	34.0	27.9	28.4	32.5	28.1	26.7
265-340 (diesel oil)	40.9	39.7	33.1	40.6	40.3	41.4
340-392 (fuel oil)	13.1	12.5	13.9	12.0	15.9	16.2
392-495 (heavy oil)	1.1	2.8	3.6	1.1	4.6	3.5

a. Figures in weight %, based on distillation curve to FBP 495°C.

B.   0.2 Pounds Coal/Hour Flow System

The process conditions for all flow reactor runs, the results of the first-step product filtration, the solvent extraction analysis, and the product gas analysis are summarized for all four runs in Table 10.

TABLE 10
0.2 Lb Coal/Hr Flow System

Reactor Packing	First-Step		Second-Step	
	Alpha-alumina	Silica	Co/Mo on SiO_2/Al_2O_3	
Reactor vol., ml	270	270	270[a]	270[b]
Temp., °C	440	440	380	380
Press., psig	1,500	1,500	1,500	1,500
H_2, scfh	7	7	7	7
Feed, g/hr	246	374	250	350
Coal/oil ratio	1:3	1:3	---	---
Run time, hrs	15.75	15.75	10.5	6.08

Product Analysis

Filtered residue, pct.	3.64	5.55	---	---
Calcd. ash, pct.	4.88	4.88	---	---
Coal dissolved, pct.	100	97	---	---

	Filtered		As received	
Benzene insolubles, %	5.1	12.4	4.5	4.4
Asphaltenes, pct.	10.6	12.4	2.3	7.7
Oil, pct.	84.3	75.2	93.2	87.9

Hydrocarbon Gas Yields, scf/lb Coal

Methane	0.638	0.140	0.025	0.189
Ethane	0.150	0.043	0.017	0.041
Propane	0.047	0.018	0.025	0.038
Butanes	0.023	0.006	0.005	0.002
TOTAL	0.858	0.207	0.072	0.268

a. Filtered product from alpha-alumina run as feed.
b. Filtered product from silica run as feed.

The residue filtered off from the first-step products, after washing with benzene to remove adsorbed oil, essentially e-qualled the calculated mineral matter in the coal feed, demonstrating essentially complete dissolution of the organic part of the coal. The benzene insolubles in the filtered product from the silica balls was more than twice that from the alpha-alumina pellets.

The total yield of total hydrocarbon gases from the alpha alumina bed reactor was nearly four times that from the silica bed reactor. Conversely, there was a nearly fourfold greater yield of hydrocarbon gases from the catalytic reactor, using the silica bed product as feed. Thus, most of the hydrocracking to give gas occurs in the alpha-alumina bed, where the concomitant carbon deposit formation does little harm, whereas a little over half of the gas formation occurs in the catalytic bed used after the silica bed, where carbon deposits are undesirable. The conversion of benzene insolubles to asphaltenes in this catalytic bed may be undesirable for similar reasons. The residence time for the liquid phase in the first-step reactor was estimated to be less than 20 minutes, compared to a residence time (not the time of contact with the packing in the baskets) of three hours in the stirred batch reactor. This probably explains the considerably lower yield of hydrocarbon gases in the flow system compared to the batch system.

In line with the larger amounts of coal-derived oil from the alpha-alumina bed, compared to the silica bed, somewhat larger amounts of identifiable coal-derived compounds were also obtained in the first-step product, as shown in Table 11, with the compounds arranged according to increasing boiling point. The alkylated compounds in the first-step product from alpha-alumina were about the same as those from silica. Thus, the greater yields of hydrocarbon gas with alpha-alumina are not due to increased dealkylation. It should be noted that some of the compounds identified were both coal-derived and alkylated.

The identities and amounts of the ten identifiable hydro-aromatics, or hydrogen donors, in the hydrogenated tar oil used to make up the slurry for the flow system are shown in Table 8. There was a substantially greater consumption of hydrogen donors in the alpha-alumina bed than in the silica bed, in line with the better quality first-step product from the alpha-alumina. The proportion of hydrogen donors consumed in the flow system alpha-alumina first-step run was actually greater than that in the corresponding batch system first-step run because of the higher ratio of tar oil to coal in the flow system, as well as the higher initial concentration of hydrogen donors.

Examinations of the used first-step packings showed a

TABLE 11
Coal-Derived Compounds and Alkylated Compounds in First-Step
Reactor[a]

First-step packing	alpha-alumina	silica
Trimethylbenzenes	0.33	0.15
Methyl-, Dimethyl-, and Ethylnaphthalenes	12.98	13.09
Methylbiphenyls	1.29	1.16
Methyl- and Dimethyldibenzofurans	4.97	3.95
Methylfluorenes[b]	0.30	0.25
Methyl- and Dimethylphenanthrenes	3.60	3.21
Methylpyrenes[b]	0.55	0.94
Benz(c)phenanthrene[b]	0.13	0.37
Methylchrysene[b]	0.32	0.16
Benz(b,j,k)fluoranthene[b]	0.57	0.21
Benz(a & e)pyrene[b]	0.28	0.07
Methylbenz(a & e)pyrene[b]	0.09	0.00
Total coal-derived compounds	2.24	2.00
Total alkylated compounds	24.43	22.91

a. Figures in weight %, calculated on the basis of the oil in the coal/oil slurry.
b. Compounds not detected in the hydrogenated tar oil used for slurry, hence derived from coal.

much greater degree of carbon deposition on the alpha-alumina than on the silica. Even though the silica packing has a much greater surface area, all of the analytical results on the products indicated greater activity with the alpha-alumina. The structure of the alpha-alumina surface may be such as to allow or promote the type of carbonaceous deposit promoting the desired hydropyrolytic reactions of the coal/hydrogenated tar oil slurry.

Scanning electron microscopy showed random stacks of flat crystallites of carbon. It is suggested that the carbon level reaches an equilibrium due to reduction by hydrogen, and physical attrition. Small amounts of carbonaceous material were found in the benzene-washed residue from filtration of the first-step products.

The filtered first-step product from the flow sytem using

TABLE 12

Densities of Slurry Vehicle Oils Compared with Densities of Filtered First-Step Products from Stirred Batch Reactor and Flow Reactor Runs[a,b]

Type reactor	Batch				Flow	
Reactor conditions	1,500 psig, 430°C				1,500 psig, 440°C	
Reactor packing	Lot B	Denstone	Alpha-alumina	Lot B-F	Silica	Alpha-alumina
Hydrogenated tar oil	1.0589	1.1000	1.0824	1.0056	1.1133	1.0600
Density increase		0.0411	0.0235		0.1077	0.0544

a. Density, at 20°C.
b. Reference (14), Table 3.

alpha-alumina packing at 1,500 psig and 440°C had an appreciably lower density than the corresponding product using silica packing (Table 12). This was not due to less coal dissolved (Table 10), but probably was because of the smaller amounts of benzene insolubles and asphaltenes in the product. As shown in Table 12, the filtered first-step product from the batch system using alpha-alumina packing at 1,500 psig and 430°C also has an appreciably lower density than the corresponding product using Denstone vitrified ceramic. For both the batch and flow systems the density increase of the product over that of the slurry vehicle oil was only about half as great for alpha-alumina as for the silica, or Denstone.

Because the first-step reactor did not contain any material capable of catalyzing extensive hydrodesulfurization or hydrodenitrification reactions, the filtered first-step products from both the batch and flow systems, with vitrified ceramic, silica, and alpha-alumina packings, all showed retention of heteroatoms (Table 13). There was essentially no reduction in nitrogen content. Identifiable nitrogen compounds included pyridines, quinolines, and carbazoles. Alpha-alumina packing in the batch system was slightly better than the vitrified ceramic for reduction of heteroatoms, and this, along with the much greater yield of hydrocarbon gases and oils, required a much greater consumption of hydrogen from hydroaromatics. Thus, in the first-step product from the alpha-alumina bed there was a little less hydrogen because of loss as hydrogen sulfide, ammonia, water, hydrocarbon gases, and volatile light oils. Silica packing was slightly better than alpha-alumina for reduction of heteroatoms in the flow system, but much inferior for the production of hydrocarbon gases and oils, so that the alpha-alumina product showed a greater consumption of hydrogen donors, and a slight lowering of hydrogen content thereby.

Conversely, the alpha-alumina bed first-step product, when processed at 1,500 psig hydrogen and 380°C in the second-step cobalt molybdate bed, gave a second-step product with fewer heteroatoms, a higher hydrogen content and a higher atomic hydrogen to carbon ratio than that from the first-step product from the silica bed (Table 14). The second-step catalyst was more active in converting first-step product from the alpha-alumina bed. A greater second-step gas yield (Table 10) with its concomitant greater carbon deposition on the catalyst was observed in converting first-step product from the silica bed. The second-step product, starting with alpha-alumina for first-step packing, had three times the increase in hydrogen content, and atomic H/C, and three times the decrease in nitrogen content as compared to the second-step product starting with silica for first-step packing.

TABLE 13
Elemental Analysis of Filtered First-Step Products from Batch and Flow Systems

Type reactor	Batch 1,800 psig, 450°C			Flow 1,500 psig, 440°C		
Reactor conditions						
Reactor packing		Denstone	Alpha-alumina		Silica	Alpha-alumina
Coal/oil slurry[b]	Lot A			Lot B-F		
Sulfur	1.70	0.37	0.30	1.29	0.16	0.37
Nitrogen	0.76	0.78	0.71	0.52	0.40	0.52
Oxygen	3.77	0.55	0.32	2.81	1.58	1.78
Carbon	86.38	90.98	92.17	87.03	90.05	89.91
Hydrogen	6.37	7.07	6.17	7.61	7.71	7.28
	98.98	99.75	100.21	99.26	99.90	99.86
Donors reacted[c]		18.2	35.8		27.5	32.6

a. Elemental weight %, all elements by analysis.
b. Moisture and ash free basis.
c. Weight percent hydrogen donors reacted.

41

TABLE 14
Flow System Second-Step Products, Elemental Analyses, Weight Percent[a]

Second-step Reactor conditions First-step packing	Hydrogenated tar oil Lot B-F	Coal/oil slurry[b] 1:3	Cobalt molybdate 1,500 psig, 380°C	
			Silica	Alpha-alumina
Sulfur	0.09	1.29	0.20	0.13
Nitrogen	0.35	0.52	0.30	0.18
Percent decrease[c]			14.3	48.6
Oxygen	0.90	2.81	1.17	1.00
Hydrogen	8.18	7.61	8.44	8.97
Percent increase[c]			3.2	9.7
Carbon	90.03	87.03	89.56	89.53
Atomic H/C	1.09	1.05	1.13	1.20
Percent increase[c]			3.7	10.1

a. Reference (14), Table 4.
b. Coal/hydrogenated tar oil, m.a.f. basis.
c. Percent change of second-step product compared to hydrogenated tar oil.

IX.   CONCLUSIONS

In the two-step coal hydroliquefaction process studied there were large differences in the behavior between various first-step nominally non-catalytic reactor packings. These results may be due to differences in the ability of the packing surface to promote or allow a sufficient amount of the type of carbonaceous deposit on which desired hydropyrolytic conversions of coal and/or asphaltenes molecules could occur by interaction with hydrogen donors (and hydrogen). The preferred packing produced asphaltenes with higher atomic H/C ratios and/or increased conversion to oil. These products appeared less likely to deactivate the cobalt molybdate catalyst used in the second-step reactor. The greater activity of the catalyst was indicated by the production of a second-step product with lower heteroatom content and higher atomic H/C, with negligible hydrocarbons gas formation and concomitant carbon deposition.

The preferred first-step packing was shown to be a moderate surface area, large pore size, very low silica content alpha-alumina. A high surface area, large pore size, pure silica, and a very low surface area, neglibible porosity vitrified ceramic were less effective. The chemical as well as the surface properties of the packing appeared to affect performance. The alpha-alumina produced more first-step hydrocarbon gas, without increasing dealkylation, giving more first step light oil, more total oil of a lower density, and less asphaltenes, along with a greater consumption of hydrogen from hydrogen donors.

V.   ACKNOWLEDGEMENTS

The authors wish to thank R. E. Lynch, D. M. Cantis, E. P. Fisher, and E. N. Eisentrout for work on reactors runs, and C. C. Kyle, H. D. Schultz, and A. L. Hiser for work on product characterization.

VI.   REFERENCES

1.   Burke, D.P., *Chem. Week* 115, 38 (1974).
2.   Klass, D.L., *Chem. Tech.* 499, (1975).
3.   Schmid, B.K., and Beuther, H., *Ind. Eng. Chem., Proc. Res. Devel.* 6, 207 (1967).
4.   Berti, V., Padovani, C. and Todesca F., *6th World Petrol. Congr.*, Sect. III, Paper 8, Frankfort (1963).
5.   Kroenig, W., "The Catalytic Hydrogenation of Coal, Tar, and Petroleum," pp. 79, 80, Springer-Verlag, Berlin, 1950.

6. Vlugter, J.C., and Van't Spijker, P., <u>8th World Petrol.</u> <u>Congr.</u>, PD 12, Paper 5, Moscow (1971).

7. Given, P.H., Cronauer, D.C., Spackman, W., Lovell, H.L., Davis, A. and Biswas, B., <u>Fuel</u> 54, 34 (1975).

8. Kovach, S.M. and Bennett, J., <u>169th National Meeting,</u> <u>ACS, Div. Fuel Chemistry, Preprints</u> 20, No. 1, 143 (1975).

9. Akhtar, S., Friedman, S. and Yavorsky, P.M., <u>U.S. BuMines</u> <u>TPR</u> 35, 11 pp. (1971)

10. Haensel, V., Pollitzer, E.L. and Watkins, C.H., <u>6th World</u> <u>Petrol. Congr.</u>, Sect. III, Paper 17, Frankfort (1963).

11. Karr, C., Jr., Mapstone, J.O., Jr., Little, L.R., Lynch, R.E. and Comberiati, J.R., <u>Ind. Eng. Chem., Product Res.</u> <u>Devel.</u> 10, 204 (1971).

12. Brahmachari, B.B., Saha, M.R., Choudhury, P.B. and Ganguli, A.K., <u>Proc. Symp. Chem. Oil Coal 1969</u>, 170 (1972).

13. Karr, C., Jr., Comberiati, J.R., McCaskill, K.B. and Estep, P.A., <u>J. Appl. Chem.</u> 16, 22 (1966).

14. Karr, C., Jr., and Abel, W.T., in "The Proceedings of the Coal Processing and Conversion Symposium", West Virginia Geological and Economic Survey, Morgantown, WV (1976).

EFFECT OF COAL MINERALS ON REACTION
RATES DURING COAL LIQUEFACTION

Arthur R. Tarrer, James A. Guin, Wallace S. Pitts
John P. Henley, John W. Prather, and Gary A Styles
Department of Chemical Engineering
Auburn University

Coal minerals represent a readily available, abundant, inexpensive source for catalytic agents for use in accelerating liquefaction and hydrodesulfurization reactions in coal conversion processes. Experimental evidence of the catalytic effect of coal minerals on hydrogenation has been reported.[1] In fact, there is a patented coal conversion process in which mineral residue is recycled because of its catalytic activity.[2] Yet the benefits of coal mineral catalysis have not been well established. The purpose of this paper is to demonstrate that certain coal minerals catalyze the hydrogenation and hydrodesulfurization of creosote oil, a coal-derived solvent used as a start-up solvent in the solvent refined coal (SRC) process; to show that, by accelerating hydrogenation of process solvent such as creosote oil, coal minerals catalysis accelerates indirectly the rate of liquefaction of coal solids; and to provide better insight as to the process advantages and disadvantages of coal mineral catalysis - more specifically, removal of coal minerals prior to hydrogenation/hydrodesulfurization, or recycle of coal mineral residue.

I. EXPERIMENTAL

A. Reagents and Materials

Creosote oil (Table 1) used in these experiments was obtained from Southern Services, Inc., and is used as a start-up solvent at the SRC pilot plant located at Wilsonville, Alabama. Southern Services, Inc., obtained the oil, creosote

TABLE 1
Gas Chromatographic Analysis of Creosote Oil

Compound	Weight %
coumarone	.10
p-/cymene	.02
indan	.11
phenol	.12
o-cresol	.05
benzonitrile	.12
p-cresol	.37
m-cresol	.16
o-ethylaniline	.03
naphthalene	5.1
thianaphthene	.08
quinoline	.37
2-methylnaphthalene	1.3
isoquinoline	.30
1-methylnaphthalene	.38
4-indanol	.55
2-methylquinoline	.42
indole	.21
diphenyl	.49
1,6-dimethylnaphthalene	.39
2,3-dimethylnaphthalene	.19
acenaphthene	6.0
dibenzofuran	6.7
fluorene	10.3
1-naphthanitrile	.18
3-methyldiphenylene oxide	1.7
2-naphthonitrile	2.4
9,10-dihydroanthracene	2.4
2-methylfuorene	.85
diphenylene sulfide	.52
phenanthrene	18.6
anthracene	4.3
acridine	.19
3-methylphenanthrene	.98
carbazole	2.2
4,5-methylenephenanthrene	2.5
2-methylanthracene	.24
9-methylanthracene	1.2
2-methylcarbazole	1.7
fluoranthene	.96
1,2-benzodiphenylene oxide	.96
pyrene	2.6

oil 24-CB, from the Allied Chemical Company. The oil has a carbon-to-hydrogen ratio of 1.25 (90.72% C and 6.05% H), a specific gravity of 1.10 at 25°C, and a boiling point range of 175° to 350°C. Kentucky No. 9/14 coal mixture was crushed; and the -170 mesh fraction - having the screen analysis shown in Table 2, and the elemental analysis in Table 3 - was used in the experiments. Table 4 lists specifications of the individual coal minerals studies. Hydrogen and nitrogen gases were the 6000 psi grade supplied by Linde. All coal was dried overnight at 100°C and 25 inches Hg vacuum before use.

B. Procedures

Basically four different types of experiments were performed: 1) catalyst screening, 2) recycle of mineral residue, 3) hydrogenation and hydrodesulfurization of demineralized coal, 4) hydrogenation and hydrodesulfurization using prehydrogenated solvent.

1. *Catalyst Screening*
Catalyst or mineral preparation consisted of grinding, followed by screening to the respective size. Depending on the hardness of the catalyst, either a diamond grinder and/or a morter and pestle was used. For each run: the charge consisted of 15 gms of catalyst, 100 gms of creosote oil, and an initial hydrogen atmostphere of 3000 psig.; reaction was carried out for two hours at 425°C and a stirrer setting of 2000 rpm. A heat-up rate of about 12 to 20°C per minute was used - requiring only about three minutes for heat-up time of about 30-35 minutes. Prior to heat-up 400 psig of hydrogen was charged to the reactor (a 300 cc magnedrive autoclave from Autoclave Engineers, Inc.) and at reaction temperature more hydrogen was added to attain the desired initial hydrogen pressure of 3000 psig. Reaction temperature (425°C) was held constant within ±3°C.

Throughout each run total pressure was recorded periodically (Figure 1); and after exactly two hours of reaction, a gas sample was collected, and the autoclave contents were quenched to below 200°C within five minutes. After allowing the catalyst to settle for one hour, a liquid sample was collected for sulfur analysis.

Between consecutive catalysts screening runs, blank runs, having no catalyst present, were made to eliminate any "memory effect." As shown in Figure 2 about three blank runs were required following a run made with the Co-Mo-Al catalyst, which - having the highest catalytic activity of those agents considered - exerted the strongest memory effect.

TABLE 2
Screen Analysis of Bituminous Kentucky No. 9/14 Coal Mixture
==

Mesh Size of Screen	% Retention
170	1.23
200	1.92
230	1.09
270	4.30
325	17.94
400	10.86
-400	62.65
Total	99.99

TABLE 3
Chemical Analysis of Bituminous Kentucky No. 9/14 Coal Mixture
==

H	4.9
C	67.8
Total Sulfur	2.55
Organic Sulfur	1.63
FeS_2	0.79
Sulfate Sulfur	0.13
Total Ash	7.16

TABLE 4
Description of Coal Minerals or Catalytic Agents Studied

Species	Classification[9]	Description[10,11]
Ankerite (Ferriferrous Dolomite)*	Carbonate	An isomorphous mixture of CaMg $(CO_3)_2$ and Ca Fe $(CO_3)_2$.
Calcite*	Carbonate	A crystalline form (hexagonal scalenohral class of the hexagonal system) of $CaCo_3$. Often, to a small extent, the Ca is replaced by iron, magnesium, and managanese. Clay, sand, bitumen, and other mechanical impurities may be present.
Dolomite*	Carbonate	A double salt with equal molecular quantities of $CaCO_3$ and $MgCO_3$, and not an isomorphous mixture of these two compounds. Usually found in a curved rhombohedral form.
Kaolinite*	Kaolin	A common type of clay; often found in minute pseudohexagonal (monoclinic) crystals. Chemically, an acid aluminum silicate, $H_4Al_2Si_2O_9$ or $2H_2O \cdot Al_2O_3 \cdot 25iO_2$ ($H_2O=14\%$). Iron is often present in small amounts.

Muscovite*	Shale	A lamina type silica substance, having a monoclinic crystal structure; and chemically classified as an acid potassium aluminum orthosilicate, $H_2KAl_3(SiO_4)_3$ or $2H_2O \cdot K_2O \cdot 3Al_2O_3 \cdot 6SiO_2(H_2O=4.5\%)$. Often, the potassium is partially replaced by sodium, and some varieties contain an excess of silicon over that indicated above.
Pyrite	Sulfide	A cubic structure of FeS_2, having, in its crystalline structure, a rocksalt-type of arrangement of Fe^{2+} and S_2^- ions, with iron being octahedrally surrounded by 5 and each S atom having one S and three Fe atoms as neighbors. Uncommonly, Ni, Co, or sometimes both are found substituted for Fe. Obtained from Matheson, Coleman, and Bell (90-95% pure).
Quartz*	Accessory	A crystallin form of SiO_2; a member of the triangonal trapezohedral class of the hexagonal system.
Siderite*	Carbonate	A crystalline form of $FeCO_3$, with the brown to gray crystals usually being rhombohedral. Calcium, mangesium, and manganese are usually present in small amounts as replacing elements.
Iron	-	Reagent grade hydrogen reduced iron from Mallinckrodt, Inc.
Reduced Pyrite	-	Solid residue from hydrogenation-of-creosote oil in presence of 15% by weight of FeS_2 at 425°C, stirrer setting of 2000 rpm, and 3000 psig of initial hydrogen pressure.
SRC Residue	-	Obtained from filter cake from Wilsonville SRC pilot plant. Analysis: 55.2% ash content and 13.6% S for -325 mesh material; and, prior to screening, 30% filter aid, 53.6% ash, and 2.9% S.
Coal Ash	-	Obtained by burning Kentucky No. 9/14 mixture (7.2% ash) in a muffle furnace at -1000°C; analysis: 13.7% iron.
Kaolin	-	Obtained from W. H. Curtis and Co.

*Minerals obtained from David New, Minerals and Books, Providence, Utah.

TABLE 5
Final Gas Analysis and Conversion for Recycle of Mineral Residue Runs

Run	Total Pressure (PSIA)	Partial Pressure (PSIA)						Conversion Based on Cresol Solubles (MMF)
		H_2	H_2S	CO_2	CH_4	C_2	C_3-C_5	
Initial	1221	825	36.9	89.5	136.6	48.9	28.7	96.3
Duplicate of Initial	1218	832	36.2	83.4	133.0	47.0	26.5	--
1st Recycle	1089	648	45.3	116.7	173.6	60.5	38.1	100.0
2nd Recycle	1064	607	45.3	113.9	179.4	22.3	33.3	100.6

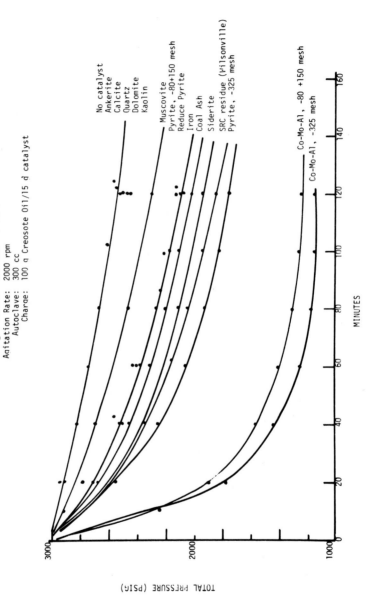

FIGURE 1. Catalytic effect of coal mineral matter as indicated by total pressure.

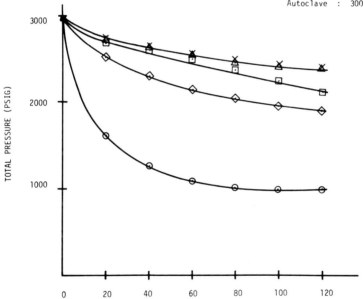

FIGURE 2. Determination of number of blank runs required to eliminate memory effect.

2. Recycle of Mineral Residue

The reaction conditions used for all of these runs were 400°C, a stirrer setting of 2000 rpm, and an initial hydrogen pressure of 2000 psig. A 3:1 solvent-to-coal weight ratio (40 gm. of coal, 120 gm. of creosote oil) was used. Two runs were made to establish a base-line for comparison. Once completed, two more runs, each having a charge with a higher concentration of mineral matter, were made: In the first, solid residue from the run with a higher mineral matter concentration was added, increasing further the mineral matter concentration. For each run, total pressure was periodically monitored (Figure 3); and final hydrogen partial pressure was measured. Final cresol soluble yields, y, were also determined (Table 5) where:

$$y = \frac{C - (R - A)}{C (1 - M)} \times 100$$

Here C is the charge of moisture free coal; R, the recovered insoluble residue; M, the fraction of mineral matter in dry coal (0.12 for Kentukcy No. 9/14 mixture); A, the mass of residue added.

To insure that solid residue was free of solvent prior to its use, after being filtered from the reaction mixture, it was washed with hot cresol and benzene, with clear benzene passing through the filter in the final wash.

C. Hydrogenation and Hydrodesulfurization of Demineralized Coal

Coal was slurried with water and partially demineralized by passing it through a high intensity magnetic separator - reducing its ash content by 64 per cent (as determined by ASTM D-271) and its total sulfur content by 25 per cent as determined by a Leco sulfur analyzer. The partially demineralized coal was then dried overnight under 25 inches Hg vacuum at 100°C; slurried with recycle solvent in a 3:1 solvent-to-coal proportion; and reacted at 410°C, 1000 psig of initial hydrogen pressure, and a 1000 rpm stirrer setting for reaction times of 15, 30, 60, and 120 minutes. At the end of each reaction, a liquid sample of reaction product was collected; the total sulfur content and cresol soluble yield (Figure 4b) was determined using a Leco sulfur analyzer and Soxhlet extraction, respectively. Assuming, on the basis of prior experimental verification[4], that the pyritic sulfur content (as determined by ASTM D2492-68) was reduced to the sulfide form (Fe_7S_8)[8] within fifteen minutes of reaction, the final

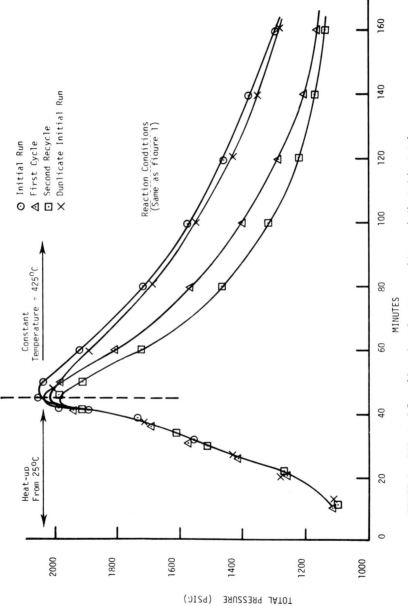

FIGURE 3. Effect of Recycling mineral matter residue as indicated by total pressure.

53

organic sulfur content of each reaction mixture was computed
(Figure 4a).

As a basis for comparison, a sample of the feed to the
magnetic separator was collected. The coal in the sample
was separated from the water by filtering; dried and reacted
in the same manner as the demineralized coal. For complete-
ness, coal that had not been exposed to water (as in the slurry feed tank to the magnetic separator) was also dried and
reacted (Figure 4).

D. Hydrogenation and Hydrodesulfurization Using Prehydro-genated Solvent

To improve the hydrogen donor activity of the solvent,
it was hydrogenated at 410°C for one hour in the presence of
15 per cent by weight of minus 150 mesh Co-Mo-Al catalyst
(Comax-451, Laporte Industries) and an initial hydrogen pressure of 2500 psig. The hydrogenated solvent was then allowed
to settle for 24 hours and doubly filtered to remove all the
Co-Mo-Al catalysts: emmission spectrophometric analysis, and
also, outside analysis by Galbraith Laboratories, Inc., showed
the Co and Mo content in the resulting hydrogenated solvent
to be less than 1ppm and 10ppm, respectively. The hydrogenated solvent has a specific gravity of 1.05 at 25°C and a carbon-to-hydrogen ratio of 1.15 (91.56% C and 6.65% H). Comparative runs were then made in which hydrogenated solvent and
untreated solvent were each reacted in a 3:1 solvent-to-coal
ratio at 410°C for 15 minutes in the presence of a nitrogen
pressure of 2000 psig, and also, in an initial hydrogen pressure of 2000 psig (Table 6).

II.   RESULTS AND DISCUSSION

Using total pressure as a rough indicator of reaction
rate, from Figure 1 some of the coal minerals definitely
appear to provide catalysis for hydrogenation of the creosote
oil. The upper and lower curves in Figure 1 represent the
extreme behavior present with no catalyst and with a commercial Co-Mo-Al catalyst, respectively. The different mineral
matter additives show evidence of catalytic activity, intermediate between these two extremes. Most interestingly, one
of the more active catalysts is filter cake residue from the
Wilsonville SRC pilot plant. Also, the catalytic activity
of -325 mesh pyrite is higher than that of -80 +150 mesh
pyrite - demonstrating that not only the composition of the
mineral matter, but also its physical state, is of considerable importance in process applications.

FIGURE 4. Effect of Demineralizing Coal Feed and Slurrying Coal
          Feed with Water on Conversion

Temperature Reaction Conditions: 410°C
           H₂ Pressure : 2000 psia @ 410°C
           Agitation Rate : 1000 rpm
           Autoclave: 300cc

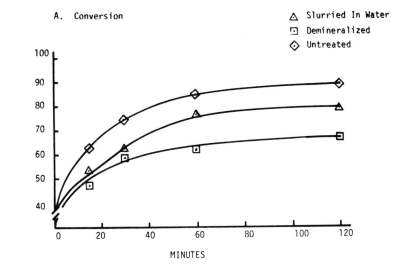

A. Conversion

△ Slurried In Water
▣ Demineralized
◇ Untreated

PERCENT CONVERSION (MAF)

MINUTES

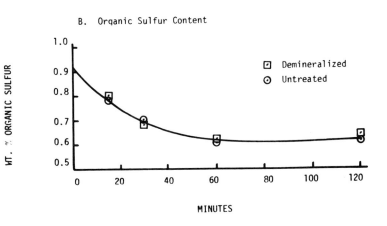

B. Organic Sulfur Content

▣ Demineralized
⊙ Untreated

WT. % ORGANIC SULFUR

MINUTES

TABLE 6
Comparison of Hydrogenation and Hydrodesulfurization Activity
of Coal in Prehydrogenated Solvent and That in Untreated
Solvent

A	B	C	D	E	F	G	H
None	410	2000psi N$_2$	15	3/1	42.13	.904 $\pm$ .023	0.34
Hydrogenation	410	2000psi N$_2$	15	3/1	83.10	.487 $\pm$ .030	.39
None	410	2000psi H$_2$	15	3/1	61.03	.912 $\pm$ .054	.35
Hydrogenation	410	2000psi H$_2$	15	3/1	90.70	.506 $\pm$ .026	.41

A. Pretreatment of Solvent
B. Temperature (^{o}C)
C. Atmosphere
D. Reaction Time (min.)
E. Solvent-to-Coal Ratio
F. Conversion (Based on Cresol Solubles)
G. Total Sulfur (%)
H. Organic Sulfur (%)

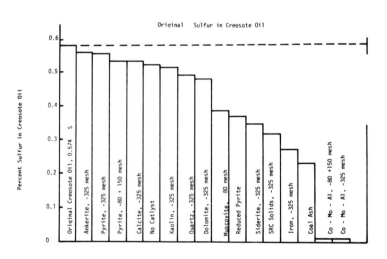

FIGURE 5. Comparison of Desulfurization Activity of Catalyst

Sulfur removal data for each of the catalyst screening runs are presented in Figure 5 and are in general agreement with the catalytic activity sequence evidenced by the total pressure data with two exceptions: pyrite, despite its pronounced effect on total pressure, appears to be a relatively poor catalyst for hydrodesulfurization. High pressure liquid chromatographic analysis of the creosote oil after hydrogenation reveals that the concentration of dibenzothiophene, an organic sulfur constituent, decreases from 1.271 + 0.03 to only 0.720 + 0.09 per cent when pyrite is present; whereas it is reduced to 0.888 + .05 when no catalyst is present and to only trace amounts ($< 0.04\%$) when Co-Mo-Al is present.[5] As stated earlier in the experimental section, pyrite is reduced rapidly during hydrogenation to the sulfide form (Fe_7S_8);[4,8] some back-reaction by the H_2S generated during reduction of the pyrite may occur, and this reaction may be partly the reason why the presence of pyrite had such a poor effect on hydrodesulfurization of the creosote oil.[12] Secondly, iron, which had a much less effect on total pressure than that of pyrite, is second only to Co-Mo-Al in sulfur removal. However, the role of iron in sulfur removal during hydrogenation is probably more as a reactant than as a catalyst, in that it reacts with any H_2S produced or directly with sulfur in the oil to form sulfides. In fact, gas analysis showed little, or no H_2S to be formed during hydrogenation of the creosote oil in the presence of iron.

An indication of hydrogenation activity is shown in Figure 6 where the final hydrogen partial pressure is presented for each of the catalyst screening runs, as determined from

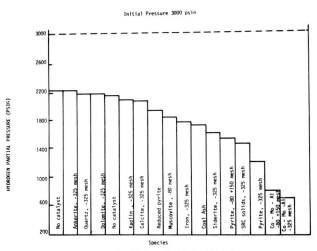

FIGURE 6. Comparison of Hydrogenation Activity of Catalyst.

gas analysis and total pressure. Again, the Co-Mo-Al is most
effective for hydrogenation; however, iron pyrite and SRC solids residue also indicate relatively high activity. Prather
et. al.[5] show, using high pressure liquid chromatography, that
the total concentration of the four major constituents in the
creosote oil - naphthalene, acenaphthene, phenanthrene, anthracene - decreases the same during hydrogenation in the presence of pyrite as it does in the presence of Co-Mo-Al, and
22 per cent more than it does when no catalyst is present.

The results of experiments showing the effect of recycling filtered mineral matter from successive autoclave runs
are shown in Figure 3; again total pressure is assumed to act
as a rough indicator of reaction rates. Obviously continued
recycle (higher concentrations) of mineral matter residue
leads to increased reaction rates, as evidenced also by
the resulting higher yields and decreasingly lower final hydrogen partial pressures (Table 5).

Further evidence that coal minerals catalyze liquefaction
reactions are given in Figure 4, in that the rate of conversion for demineralized coal is much slower than that of untreated coal. In addition, soaking of the coal in water, or slurrying with water, causes also a significant decrease in the
rate of liquefaction. Some of the coal minerals - particularly sulfates - are soluble in water, and thus, are extracted by
soaking the coal in water, as evidenced by the 0.12 per cent
decrease in total sulfur content of the coal with soaking (Table 7), which is about the same as the per cent sulfur (0.13%)
present in the sulfate form in the untreated coal. Yet since

TABLE 7
Ash and Total Sulfur Content of Kentucky No. 9/14 Coal and
That After Slurrying with Water and After Deminerization

	Ash Content (%) (ASTM D-27)	Total Sulfur Content (%)	Conversion* After Two Hours of Reaction
Original Coal	7.16	2.55	87.8
Coal from Slurry Feed Tank	6.28	2.43	77.38
Demineralized Coal	2.55	1.90	67.23

*Refer to Figure 4

exposure of the coal to water may affect the chemical characteristics of the coal in various ways other than removal of
soluble minerals, further experimental study is needed to determine conclusively why slurrying coal with water prior to
hydrogenation decreases its rate of liquefaction.

Despite the significant effect of demineralization of the
coal on its liquefaction behavior, its organic hydrodesulfurization activity remained practically the same (Figure 4a).
Assuming that mostly pyrite was removed by the magnetic separator, then no significant difference in the organic hydrodesulfurization activity of the demineralized coal and that of untreated coal should result, and the results given in Figure
4a should be expected; for, as shown in Figure 5, pyrite has
relatively little overall catalytic effect on hydrodesulfurization of creosote oil. Depending then on the composition of
coal minerals - eg. high pyrite content, etc. - the relative
effect of coal mineral catalysis can be significantly greater
for liquefaction than for organic hydrodesulfurization. As a
result, coal mineral catalysis during hydrogenation/hydrode-
sulfurization of coal may or may not be advantageous, depending on process objectives and on composition of the coal minerals. To illustrate: if hydrogenation is limiting, for
example, as may be the case in producing a synthetic fuel
oil, then catalysis by coal minerals of hydrogenation reactions would be advantageous; and thus, so would recycle of
coal minerals. On the other hand, if hydrodesulfurization is
limiting and, as usual, minimum hydrogenation is desired,
which is often the case in SRC and related processes, removal
of coal minerals such as pyrite prior to hydrogenation/hydro-
desulfurization would be advantageous; for, to attain the required amount of sulfur removal, more hydrogenation would
occur when all the catalytic coal minerals are present than
when no pyrite, and similar behaving catalytic coal minerals,
is present. That is, in the presence of coal minerals such
as pyrite, excess hydrogenation - more than that required to
liquefy the coal so that mineral residue can be separated by
filtration, etc. - would occur.

For coal particles to dissolve in a carrier solvent, i.e.
liquefy, at temperatures of 385 to 450ºC, either molecular
hydrogen or hydrogen donor species must be available to transfer hydrogen to the coal.[6] A direct relationship exists between the degree of dissolution and hydrogen transfer: the
more hydrogen transferred, the greater the liquefaction.[7]
Coal mineral matter, being solid in form, most likely can not
directly catalyze hydrogen transfer to coal solids either from
molecular hydrogen dissolved in the carrier solvent or from
hydrogen donor species. More reasonably, coal minerals can
catalyze transfer of dissolved molecular hydrogen to the sol-

vent - i.e. hydrogenation of the solvent. But, does hydrogenation of the solvent increase its hydrogen donor activity? If indeed it does, then the rate of liquefaction of coal slurried with prehydrogenated solvent should be greater than that of coal slurried with untreated solvent. To verify whether or not this is true, the conversion of coal solids obtained in the prehydrogenated solvent experiments were compared with those obtained with untreated solvent (Table 6). Apparently prehydrogenation of the solvent increases its hydrogen donor activity significantly, for the conversion obtained with the prehydrogenated solvent was 97 and 49 per cent higher than that obtained with untreated solvent in a nitrogen and a hydrogen atmosphere, respectively. Yet, practically the same amount of sulfur removal results when either prehydrogenated or untreated solvent are used. Apparently then, coal minerals serve to catalyze hydrogenation of the solvent, increasing its hydrogen donor activity, and thereby, the rate of hydrogen transfer to the coal, thus the rate of liquefaction.

Supportive evidence that coal minerals serve primarily to catalyze hydrogenation of the donor solvent is provided by Curran, et. al.:[1] they found that, in a nitrogen atmosphere, " . . . all attempts to accelerate hydrogen transfer to coal slurried in tetralin with contact type of catalysts of the hydrofining type (cobalt molybdate on alumina) or with cracking catalysts (silica-alumina) were unsuccessful." Whereas, in a hydrogen atmosphere - as shown here and by others[1] - the rate of liquefaction increases directly with increases in the concentration of coal minerals. The rate limiting step in liquefaction furthermore appears to be the transfer of dissolved molecular hydrogen to the donor solvent, with the transfer of hydrogen from the donor solvent to coal solids occurring rapidly.

III. CONCLUSIONS

Certain coal minerals - particularly pyrite - catalyze hydrogenation of coal-derived solvents such as creosote oil and SRC recycle solvent. The rate limiting step in liquefaction of coal is the transfer of hydrogen to donor solvent, and the rate of liquefaction increases directly with the concentration of coal minerals. Certain coal minerals also catalyze hydrodesulfurization of creosote oil - pyrite having a relatively insignificant effect on total hydrodesulfurization. The physical state, as well as chemical composition, of the coal minerals affect hydrogenation and hydrodesulfurization activity during coal liquefaction. Coal mineral catalysis of hydrogenation and hydrodesulfurization reactions occurring in coal conversion processes may or may not be advantageous, depending on process objectives and on composition of the coal minerals.

IV. ACKNOWLEDGEMENT

The research reported here is supported by the RANN division of the National Science Foundation under Grant No. 38701, by Auburn University Engineering Experiment Station, and by the Alabama Mining Institute. The authors are especially indebted to Dr. Y. A. Liu and Dr. M. C. Lin for preparing demineralized coal samples using high intensity magnetic separation, and also, to Southern Services, Inc., for supplying various materials for the reported experiments and for the helpful advice of their staff, particularly Everett L. Huffman, throughout the course of the work.

V. REFERENCES

1. Wright, C.H., and Severson, D. E., ACS Div. of Fuel Chem. Preprints 16 (2), 68, (1972).
2. Hinderliter, C. R., and Perrussel, R. E., U.S. Patent 3, 884, 796, (May 20, 1975).
3. Pitts, W. S., M.S. Thesis, Auburn University, Auburn, Alabama (1976).
4. Henley, J. P., Jr., M.S. Thesis, Auburn University, Auburn, Alabama (1975).
5. Prather, J. W., Tarrer, A. R., Guin, A. J., and Johnson, D. R., "High Pressure Liquid Chromatographic Studies of Coal Liquefaction," ACS, Div. of Fuel Preprints, This Volume, (1976).
6. Guin, J. A., Tarrer, A. R., Taylor, Z. L., and Green, S. C., ACS Div. of Fuel Chem., 20 (1), 66, (April, 1975).
7. Curran, G. P., Struck, R. T., and Gorin, E., Ind. Eng. Chem. Process Design & Develop., 6, 166, (1967).
8. Ric hardson, J. T., Fuel, 51, 150 (1972).
9. Nelson, J. B., BCURA Mon. Bull., 17, No. 2, 1953, p. 41.
10. Mason, B. H., and Berry, L. G., Elements of Minerology, 1968, W. H. Freeman and Company.
11. Rogers, A. F., Introduction to the Study of Minerals, 1936, McGraw-Hill.
12. Yergey, Alfred L., Lampe, F. W., Vestal, M. L., Day, A. G., Fergusson, G. J., Johnston, W. H., Sryderman, J. S., Essenhigh, R. H., Hudson, J. E., Ind. Eng. Chem. Process Des. and Develop. 13 233 (1974).

HYDROGENATION OF PHENANTHRENE OVER A
COMMERCIAL COBALT MOLYBDENUM SULFIDE
CATALYST UNDER SEVERE REACTION CONDITIONS

Chao-Sheng Huang, Kuo-Chao Wang, and Henry W. Haynes, Jr.
University of Mississippi

*This paper reports the results of a study of the hydro-
genation of phenanthrene over a commercial cobalt molybdenum
sulfide catalyst (Nalcomo 471) over a range of conditions
considered to be moderate to very severe: Temperature =
600-1000°F, Pressure = 1500-2500 psig, Space Velocity =
1.0-4.0 gm/hr/gm, Hydrogen Treat Rate = 5,000-15,000 scf/bbl.
The reaction was studied in a steady state flow reactor containing a five gram catalyst charge. The analysis of the
gaseous products was by gas chromatography. The liquid products were analyzed by gas chromatography-mass spectrometry.
Hydrogenation was extensive at the lower temperatures, and
cracking reactions became significant at temperatures in
excess of 800°F. Most of the cracked products were produced
by the successive saturation, ring opening, and cleavage of
terminal rings. Large quantities of perhydrophenanthrene
isomers were found in the products, and some cracking of
this species was indicated. However, at the temperatures
required for cracking, the thermodynamic equilibrium was
less favorable for perhydrophenanthrene formation. Only
trace quantities of biphenyls were observed indicating that
saturation and cleavage at the central ring did not occur
to an appreciable extent.*

I. INTRODUCTION

Cobalt molybdenum catalysts were developed primarily for
the hydrodesulfurization of petroleum residuum streams; however, they have been applied extensively in laboratory and
pilot plant investigations of the production of quality synthetic fuels from coal, oil shale and tar sands. In these
applications the catalyst has been of interest not just because of its desulfurization capabilities, but also because
of its high activity in hydrogenation, stabilization and

conversion reactions. The remarkable feature of cobalt molybdenum catalysts is their ability to remain active despite the presence of notorious catalyst poisons, in particular organic sulfur and nitrogen compounds, in the feedstocks undergoing treatment.

Phenanthrene is typical of the hydrocarbons produced during the liquefaction of coal. The staggered phenanthrene-like compounds are thermodynamically more stable (1) than the linear anthracene-like isomers and they are usually present in greater abundance in coal derived liquids (e.g. 2). Partially hydrogenated derivatives of phenanthrene are very active hydrogen donors in coal extraction. In one study 9, 10-dihydrophenanthrene was reported to be slightly superior to tetralin in hydrogen donor activity (3). Perhydrophenanthrene was much less active, and the possibility of over-hydrogenating the solvent in a hydrogen donor coal liquefaction scheme is widely recognized. The extent to which phenanthrene is hydrogenated in a catalytic solvent hydrogenation reactor is therefore of considerable interest.

In addition some conversion to lower molecular weight species is usually desirable. While cobalt molybdenum is much less active in this regard as compared with catalysts containing an acidic component, it has proven superior in terms of hydrocracking selectivity in at least one instance. Gardner and Hutchinson found cobalt molybdenum to be active and selective for hydrocracking polyphenyls including biphenyl (4). Catalysts on acidic supports were less selective and produced mostly coke. Penninger and Slotboom observed substantial quantities of 2-ethylbiphenyl and biphenyl in the reaction products from the thermal high pressure hydrogenolysis of phenanthrene indicating that hydrogenation and α-ring-opening at the 9, 10-position was in fact taking place (5). The product distribution from cracking over non-acidic or low acidity catalysts frequently resembles that obtained in thermal cracking processes. Since cobalt molybdenum catalysts are known to be selective in the hydrocracking of biphenyl, and it might be speculated that biphenyls can be formed from phenanthrene over cobalt molybdenum in a manner similar to that observed in thermal cracking, it was hoped that some cracking at the central ring of phenanthrene might be accomplished. This speculation was a major driving force behind the present investigation. The economic advantages of hydrocracking at the inner rings of condensed ring aromatics as compared with terminal ring cleavage are readily apparent in terms of reduced hydrogen consumption, higher yields and in some cases higher quality products.

It was evident from the very beginning of this investigation that much higher temperatures than normally encountered in packed bed reactors would be required in order to obtain substantial yields of cracked products. Catalyst deactivation due to carbon formation on the catalyst surface would likely be a problem. However, liquid fluidized beds have been employed on a commercial scale in the hydrodesulfurization of petroleum residuum streams (H-Oil) and on the pilot plant scale in the liquefaction of coal (H-Coal). One of the advantages of the liquid fluidized bed reactor is that provisions can be made for the continuous addition and withdrawal of catalyst. The addition of fresh catalyst could conceivably overcome the deactivation problem when operating at high severity.

II. EXPERIMENTAL

The catalyst employed in this investigation was supplied by the Nalco Chemical Company and carries the designation Nalcomo-471. According to the manufacturer's specifications the catalyst consists of 12.5% MoO_3 and 3.5% CoO supported on an alumina base. The surface area and total pore volume are 295 m^2/gm and 0.55 cc/gm respectively. High purity hydrogen (99.995% according to the supplier's specifications) was obtained from the Matheson Gas Products Company in 3500 psig cylinders. Phenanthrene, 98+% purity, melting point 99-101°C was purchased from the Aldrich Chemical Company. An elemental analysis of the phenanthrene (Galbraith Laboratories, Knoxville, TN) indicated that the sample consisted of Carbon: 93.69%, Hydrogen: 5.48%, Nitrogen: 0.01%, Sulfur: 0.42%, and Oxygen: 0.39% by weight.

The reactor, Fig. 1, is a steady flow type constructed of a 1/2 inch heavy wall (0.083) Type 316 stainless steel tube and heated by a Marshall tubular furnace, model 1016. The reactor charges approximately five grams of catalyst. Thermocouples are inserted about 1/2 inch into both ends of the catalyst bed, and a preheat zone of glass chips is provided at the bed inlet. Liquid phenanthrene is metered into the reactor by a precision Ruska proportioning pump, model 2252-BI, with a heated barrel. Various discharge rates from 2 cc/hr to 240 cc/hr can be obtained by selecting the proper choice of gear ratios. The hydrogen flowrate is monitored by a flow meter constructed of a 29 inch length of 0.009 inch I.D. capillary tubing and a Barton model 200 differential pressure cell. The capillary pressure and reactor pressure are controlled respectively by a Tescom pressure regulator, model 26-1023-002 and a Tescom back pressure regulator model 26-1723-24. Flow rates are controlled with a Hoke Milli-Mite needle valve. Liquid products are collected in two high pressure accumulators constructed of one inch schedule 80 stainless steel pipe and Swagelok buttweld

connectors. Product gases are vented through a low pressure
accumulator in dry ice-propanol, through a wet test meter,
and collected in a polyethylene gas bag.

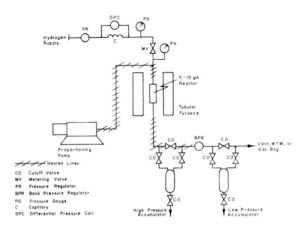

Fig. 1. Simplified Flow Diagram of Apparatus.

The catalyst was crushed and sieved to 20/30 mesh and
calcined at 1000°F in air for four hours. After calcining,
the catalyst was charged to the reactor and the system pressure tested with hydrogen. Presulfiding was carried out at
250 psig using a hydrogen sulfide (2%) - hydrogen (98%) mixture. During presulfiding the gas flow was set at 5 l(STP)
hr/gm catalyst and the reactor temperature was maintained at
400°F for 5 hours. After this period the temperature was
raised at a rate of 2°F/min to 600°F and held for 1 hour.
Then with sulfiding gas flowing at a minimal rate the reactor
was cooled to room temperature.

The phenanthrene feed was spiked with elemental sulfur
to a total sulfur content of 1.0% by weight. During startup
the spiked feed was cut in with the hydrogen sulfide-hydro-
gen mixture flowing at 1000 psig and 500°F. Once liquid
product was detected in the high pressure accumulator the gas
mixture was replaced with pure hydrogen and the reactor was
brought to operating conditions. After a period of time
sufficient for three displacements of the reactor volume the
reaction products were directed to a second high pressure
accumulator and a yield period begun. At the termination of
a yield period the liquid product was collected and stored in
a freezer, and the gas product was immediately analyzed. The
system was brought to a new set of run conditions and the procedure repeated. During all adjustments care was taken to
assure that the rate of temperature rise never exceeded

120°F/hr and that the catalyst was at all times in contact
with sulfur.

The products were analyzed on a gas chromatograph which
utilized a hydrogen flame ionization detector and possessed
temperature programming capabilities. The column for the
liquid product analysis was packed with 5% SE-30 on 60/80
mesh Chromosorb P, AW. The gas analysis column was packed
with Chromosorb 102. The identification of the various product peaks was accomplished by measuring the retention time
of pure compounds and by a GC-mass spectral analysis. The
former method was used to identify most of the lower molecular
weight hydrocarbons and the latter method was relied upon for
identification of many of the high molecular weight peaks.
The mass spectra of some of the more important product peaks
are presented in Fig. 2. Additional information on the analytical methods used in this investigation is available in
masters theses by Huang (6) and Early (7).

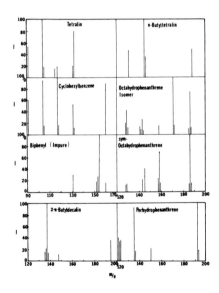

Fig. 2.   Mass Spectra of Selected Product Peaks.

III.   RESULTS

A total of eighteen yield periods was successfully completed in two series of experiments. The operating conditions
and product yields are presented in Tables 1 through 4.
These yields have been adjusted to meet a 100% carbon material
balance. Before discussing these results it must be pointed

TABLE 1

WH-08 RUN SUMMARY AND COLLECTIVE PRODUCT YIELDS

HYDROGENATION OF PHENANTHRENE OVER 5.36 GRAMS OF SULFIDED CoMo/ALUMINA CATALYST—NALCOMO 471, #74-5372

Yield period no.	1	2	3	4	5	6	7	8	9	10	11	12	13
Yield period length, hrs.	3.00	2.00	1.50	3.00	1.50	2.00	4.00	3.00	3.00	3.00	3.00	2.00	3.00
Top temperature, °F	861	902	812	947	949	945	950	861	1000	997	1003	1003	858
Bottom temperature, °F	836	897	789	945	949	948	945	846	1000	1002	998	1003	844
Pressure, psig	2000	2000	2000	2000	2000	2000	2000	2000	2000	1510	2500	2000	2000
Liquid feed rate, cc/hr	10.3	10.2	10.2	10.2	20.6	16.4	5.1	10.2	10.2	10.3	10.3	10.1	10.2
Liquid space velocity, gm/hr/gm	1.94	1.93	1.94	1.94	3.90	3.11	0.97	1.94	1.94	1.95	1.95	1.91	1.94
Space time, (gm/hr/gm)-1	0.52	0.52	0.52	0.52	0.26	0.32	1.03	0.52	0.52	0.51	0.51	0.52	0.52
Approx. H_2 treat rate,													
L(STP)/hr	15.0	29.5	31.5	17.0	28.5	27.5	12.0	16.0	16.0	12.0	20.5	15.5	14.5
L(STP)/gm	1.44	2.85	3.03	1.64	1.37	1.65	2.31	1.54	1.54	1.15	1.97	1.51	1.40
Exit gas rate, L(STP)/hr	9.7	23.3	25.3	14.1	22.0	21.7	9.8	11.9	12.1	10.9	16.4	11.2	10.3
Cum. hrs. on catalyst	3.8	7.9	11.0	19.5	23.8	27.3	32.8	38.0	43.0	50.5	55.3	60.3	67.8
Cum. gms oil/gm catalyst	7.2	13.9	18.9	33.8	42.8	46.3	52.9	59.5	67.5	71.8	79.6	87.0	96.7
Liq. Mat'l balance, wt %	104.6	100.5	101.6	91.7	95.9	97.3	90.4	90.0	83.1	90.4	87.8	83.7	106.4
Carbon Mat'l bal., wt %	101.1	104.0	100.3	99.9	101.4	104.5	105.8	89.9	100.3	99.7	98.0	103.0	104.8
CORRECTED YIELDS BASED ON LIQUID FEED													
Feed conversion, mole %	93.7	79.4	97.8	86.1	66.5	66.2	94.5	86.3	80.0	45.9	89.0	73.0	68.2
Conversion to C13-, mole %	17.9	22.3	5.2	67.3	36.7	32.0	83.8	12.2	69.8	30.7	80.1	59.1	2.72
Hydrogen consumption,													
L(STP)/gm	0.69	0.54	0.72	0.58	0.44	0.44	0.80	0.59	0.64	0.32	0.63	0.60	0.31
wt %	6.2	4.8	6.5	5.2	4.0	4.0	7.2	5.3	5.8	2.9	5.7	5.4	2.8
Gas yield (C1-C4), wt %	4.6	9.3	5.2	16.5	12.9	12.7	30.1	6.5	25.7	12.5	18.7	25.3	1.4
Liquid yield (C5+), wt %	101.5	95.6	101.3	88.7	91.1	91.3	77.2	98.9	80.1	90.4	86.9	80.1	101.4
C5-C8 yield, mole %	3.0	1.2	0.0	18.1	2.1	1.0	27.0	0.0	18.1	0.9	28.6	7.4	0.3
C9-C12 yield, mole %	16.8	19.1	0.95	60.5	33.4	28.1	61.3	8.9	53.4	26.5	68.3	45.8	1.90
C13-C14 yield, mole %	82.1	77.7	94.8	32.7	63.3	68.0	16.2	87.8	30.2	69.3	19.9	40.9	97.3

TABLE 2

WH-08 PRODUCT YIELDS

HYDROGENATION OF PHENANTHRENE OVER 5.36 GRAMS OF SULFIDED CoMo/ALUMINA CATALYST—NALCOMO 471, #74-5372

CORRECTED PRODUCT YIELDS BASED ON LIQUID FEED, MOLE %

Formula	YIELD PERIOD NO.	1	2	3	4	5	6	7	8	9	10	11	12	13
C 1H 4	Methane	2.22	6.63	4.56	17.44	11.45	13.16	32.87	4.86	52.38	26.17	32.34	46.62	1.08
C 2H 6	Ethane	2.07	5.15	3.47	14.44	8.80	10.31	23.81	4.19	38.16	18.74	26.41	34.77	1.45
C 2H 4	Ethylene	0.00	0.00	0.00	0.00	0.00	0.00	0.00	0.00	0.00	0.00	0.00	0.00	0.00
C 3H 8	Propane	2.97	7.09	3.78	15.42	11.66	12.52	27.54	5.26	30.43	16.06	20.98	34.34	1.57
C 4H10	Isobutane	0.00	0.07	0.00	0.22	0.20	0.27	0.00	0.00	0.15	0.22	0.56	0.55	0.00
C 4H10	Butane	10.33	18.45	9.98	26.37	22.73	20.15	49.94	12.30	21.26	8.93	18.40	20.11	1.97
C 5H10	Cyclopentane	0.00	0.00	0.00	0.00	0.00	0.00	0.31	0.00	0.24	0.00	0.48	0.00	0.00
C 6H14	2-Methylpentane	0.00	0.00	0.00	0.00	0.16	0.00	0.08	0.00	0.28	0.00	0.56	0.07	0.00
C 6H14	n-Hexane	0.00	0.00	0.00	1.09	0.00	0.00	3.00	0.00	1.22	0.00	2.38	0.51	0.00
C 6H 6	Benzene	0.97	0.64	0.00	3.12	0.35	0.09	4.91	0.00	3.63	0.00	3.80	0.48	0.00
C 7H16		0.00	0.00	0.00	0.98	0.00	0.00	1.48	0.00	0.23	0.00	0.73	0.00	0.00
C 7H16		0.60	0.00	0.00	1.21	0.00	0.00	1.74	0.00	0.00	0.00	1.04	0.00	0.00
C 7H 8	Toluene	0.54	0.13	0.00	4.18	0.83	0.62	5.39	0.00	4.96	0.63	7.17	2.32	0.10
C 7H16		0.00	0.00	0.00	0.53	0.18	0.20	0.20	0.00	0.00	0.00	0.26	0.00	0.00
C 8H10	Ethylbenzene	0.24	0.00	0.00	2.45	0.18	0.03	0.87	0.00	2.60	0.00	3.13	1.28	0.09
C 8H10	Xylene	0.00	0.00	0.00	0.00	0.00	0.00	1.41	0.00	0.21	0.00	0.24	0.00	0.08
C 8H10	Alkylbenzene	0.18	0.00	0.00	1.67	0.00	0.00	1.28	0.00	1.19	0.00	2.34	0.10	0.00
C 9H12	Alkylbenzene	0.17	0.00	0.00	0.59	0.00	0.00	2.14	0.00	0.30	0.00	1.42	0.00	0.00
C 8H10	p-Xylene	0.43	0.44	0.00	2.90	0.60	0.24	6.36	0.00	3.48	0.28	6.45	2.62	0.00
C10H14	n-Butylbenzene	4.23	2.58	0.00	6.53	2.46	1.61	8.33	0.90	3.89	0.10	7.05	1.98	0.00
C10H18	Decalin	1.76	0.61	0.00	5.68	2.67	2.39	8.76	0.05	5.70	0.42	8.17	4.18	0.00
C11H16	Methylbutylbenzene	0.00	0.00	0.00	0.00	0.00	0.00	1.05	0.00	0.54	0.00	1.34	0.11	0.00
C10H12	Tetralin	6.56	11.11	0.34	17.82	9.79	8.74	17.02	4.34	8.66	2.44	16.56	6.89	1.56
C10H 8	Naphthalene	0.71	4.19	0.00	22.24	11.51	10.16	17.42	1.25	27.25	16.08	25.48	21.40	0.00
C11H14	Methyltetralin	0.12	0.14	0.00	0.00	0.55	0.41	0.74	0.38	0.27	0.10	0.75	1.19	0.00
C11H10	Methylnaphthalene	0.59	0.07	0.00	0.95	1.03	0.60	1.25	0.24	0.23	0.09	1.00	0.54	0.22
C12H16	Cyclohexylbenzene	0.08	0.13	0.58	4.66	3.19	2.90	2.90	0.92	4.37	3.93	3.66	6.36	0.12
C12H16	Ethyltetralin	1.00	0.03	0.00	0.68	0.53	0.28	0.86	0.24	0.62	0.56	1.10	0.44	0.00
C12H10	Biphenyl	0.05	0.07	0.03	0.00	0.07	0.04	0.02	0.04	0.00	0.00	0.50	0.10	0.00
C12H12	Ethylnaphthalene	0.35	0.20	0.03	0.61	0.50	0.18	0.20	0.21	0.59	0.41	0.42	0.65	0.00
C14H26	n-Butyldecalin	1.15	3.84	2.88	0.74	1.12	0.83	0.62	0.56	0.93	2.31	0.88	1.93	0.08
C14H24	Perhydrophenanthrene Isomer	11.32	1.99	2.21	1.28	1.18	1.20	1.11	4.93	0.06	0.12	0.26	0.59	0.00
C14H24	Perhydrophenanthrene Isomer	4.51	1.44	20.41	1.10	0.00	1.30	0.56	2.29	0.07	0.13	0.02	0.42	0.00
C14H24	Perhydrophenanthrene Isomer	4.74	3.18	8.03	1.23	1.41	1.05	0.67	3.54	0.32	0.76	0.61	1.10	1.48
C14H24	Perhydrophenanthrene Isomer	2.45	11.10	16.57	0.03	0.89	0.69	0.00	2.00	0.03	0.00	0.00	0.00	3.76
*C14H20	+ n-Butyltetralin	22.99	0.00	0.00	2.74	3.87	3.69	1.06	20.94	0.58	0.70	0.82	1.10	15.08
C14H18 / *C14H16	Octahydrophenanthrene Isomer + n-Butylnaphthalene	12.82	16.75	10.60	6.35	11.22	12.44	4.25	14.92	5.00	6.47	4.52	5.52	0.82
C14H14	Ethylbiphenyl	1.38	2.84	2.06	1.30	3.44	4.16	0.94	1.56	1.06	1.60	0.85	1.45	6.53
C14H18 / *C14H18	Octahydrophenanthrene Isomer + Dihydrophenanthrene	3.73	3.49	0.00	1.66	2.68	3.73	0.98	4.20	1.54	3.59	0.85	1.63	0.00
C14H18	Octahydrophenanthrene Isomer	5.17	2.82	22.04	1.20	1.14	0.94	0.40	7.38	0.30	0.26	0.20	0.41	13.23
C14H14	Tetrahydrophenanthrene	6.72	9.61	7.78	1.83	3.97	5.06	0.74	12.35	1.27	1.54	0.73	1.64	24.48
C14H10	Phenanthrene	6.28	20.60	2.16	13.93	33.49	33.79	5.51	13.68	20.00	54.10	10.99	27.04	31.82
	TOTALS	119.43	135.36	117.48	185.18	153.69	153.56	238.72	123.28	244.03	166.76	215.45	230.43	105.53

*THESE COMPONENTS WERE NOT RESOLVED ON THE GAS CHROMATOGRAMS.

TABLE 3

WH-09 RUN SUMMARY AND COLLECTIVE PRODUCT YIELDS

HYDROGENATION OF PHENANTHRENE OVER 5.16 GRAMS OF SULFIDED CoMo/ALUMINA CATALYST—NALCOMO 471, #74-5372

	1	2	3	4	5
Yield period no.					
Yield Period length, hrs.	3.00	3.00	3.00	3.00	3.00
Top temperature, F	617	683	744	791	617
Bottom temperature, F	583	620	653	709	586
Pressure, psig	2000	2000	2000	2000	2000
Liquid feed rate, cc/hr	10.2	10.2	10.3	10.2	10.2
Liquid space velocity, gm/hr/gm	2.01	2.01	2.02	2.01	2.01
Space time, (gm/hr/gm)-1	0.50	0.50	0.49	0.50	0.50
Approx. H_2 treat rate,					
L(STP)/hr	13.5	14.8	13.5	14.0	13.0
L(STP)/gm	1.30	1.43	1.29	1.35	1.25
Exit gas rate, L(STP)/hr	10.8	10.3	8.2	8.1	10.9
Cum. hrs. on catalyst	3.4	8.7	13.4	18.4	23.4
Cum. gms oil/gm catalyst	9.1	17.9	25.9	34.1	42.5
Liq. mat'l balance, wt %	106.2	105.1	109.0	107.3	104.6
Carbon mat'l bal., wt %	102.0	100.1	102.9	100.4	100.5
CORRECTED YIELDS BASED ON LIQUID FEED					
Feed conversion, mole %	90.8	97.6	99.5	99.2	89.4
Conversion to C13-, mole %	1.3	1.3	1.4	1.5	1.0
Hydrogen consumption,					
L(STP)/gm	0.46	0.55	0.66	0.77	0.46
wt %	4.1	5.0	5.9	6.9	4.1
Gas yield (C1-C4), wt %	0.0	0.0	0.01	0.3	0.0
Liquid yield (C5+), wt %	104.1	105.0	105.9	106.6	104.1
C5-C8 yield, mole %	0.0	0.0	0.0	0.0	0.0
C9-C12 yield, mole %	1.5	1.5	1.6	1.5	1.2
C13-C14 yield, mole %	98.7	98.7	98.6	98.5	99.0

TABLE 4

WH-09 PRODUCT YIELDS

HYDROGENATION OF PHENANTHRENE OVER 5.16 GRAMS OF SULFIDED CoMo/ALUMINA CATALYST—NALCOMO 471, #74-5372

CORRECTED PRODUCT YIELDS BASED ON LIQUID FEED, MOLE %

	YIELD PERIOD NO.	1	2	3	4	5
C 1H 4	Methane	0.03	0.03	0.04	0.09	0.01
C 2H 6	Ethane	0.01	0.01	0.00	0.13	0.01
C 2H 4	Ethylene	0.00	0.00	0.00	0.00	0.00
C 3H 8	Propane	0.02	0.01	0.01	0.09	0.01
C 4H10	Isobutane	0.07	0.00	0.00	0.75	0.00
C 4H10	Butane	0.00	0.02	0.02	0.00	0.02
C 5H10	Cyclopentane	0.00	0.00	0.00	0.00	0.00
C 6H14	2-Methylpentane	0.00	0.00	0.00	0.00	0.00
C 6H14	n-Hexane	0.00	0.00	0.00	0.00	0.00
C 6H6	Benzene	0.00	0.00	0.00	0.00	0.00
C 7H16		0.00	0.00	0.00	0.00	0.00
C 7H16		0.00	0.00	0.00	0.00	0.00
C 7H 8	Toluene	0.00	0.00	0.00	0.00	0.00
C 7H16		0.00	0.00	0.00	0.00	0.00
C 8H10	Ethylbenzene	0.00	0.00	0.00	0.00	0.00
C 8H10	Xylene	0.00	0.00	0.00	0.00	0.00
C 8H18	Alkylbenzene	0.00	0.00	0.00	0.00	0.00
C 9H12	Alkylbenzene	0.42	0.56	0.49	0.94	0.47
C 8H10	p-Xylene	0.00	0.00	0.00	0.00	0.00
C10H14	n-Butylbenzene	0.00	0.00	0.00	0.00	0.00
C10H18	Decalin	1.05	0.90	1.09	0.05	0.68
C11H16	Methylbutylbenzene	0.00	0.00	0.00	0.00	0.00
C10H12	Tetralin	0.00	0.00	0.00	0.25	0.00
C10H 8	Naphthalene	0.00	0.00	0.00	0.13	0.00
C10H12		0.00	0.00	0.00	0.00	0.00
C11H14	Methyltetralin	0.00	0.00	0.00	0.15	0.00
C11H10	Methylnaphthalene	0.00	0.00	0.00	0.00	0.00
C12H16	Cyclohexylbenzene	0.00	0.00	0.00	0.00	0.00
C12H16	Ethyltetralin	0.00	0.00	0.00	0.00	0.00
C12H10	Biphenyl	0.00	0.00	0.00	0.00	0.00
C12H12	Ethylnaphthalene	0.00	0.00	0.00	0.00	0.00
C14H26	n-Butyldecalin	0.00	0.00	0.61	5.00	0.00
C14H24	Perhydrophenanthrene Isomer	0.00	0.88	4.06	6.28	0.00
C14H24	Perhydrophenanthrene Isomer	1.17	6.84	14.77	28.04	2.29
C14H24	Perhydrophenanthrene Isomer	0.99	5.02	12.17	16.18	0.90
C14H24	Perhydrophenanthrene Isomer	1.47	5.65	11.79	14.86	1.64
*C14H20	+ n-Butyltetralin					
C14H18	Octahydrophenanthrene Isomer	25.12	23.83	14.94	7.33	24.57
*C14H16	+ n-Butylnaphthalene					
C14H14	Ethylbiphenyl	0.00	0.00	0.00	0.00	0.00
C14H18	Octahydrophenanthrene Isomer	27.32	11.27	10.19	5.68	28.57
*C14H12	+ Dihydrophenanthrene					
C14H18	Octahydrophenanthrene Isomer	30.35	41.38	27.42	12.89	26.56
C14H14	Tetrahydrophenanthrene	3.13	1.52	2.18	1.47	3.86
C14H10	Phenanthrene	9.17	2.36	0.51	0.77	10.60
	TOTALS	100.31	100.27	100.29	101.08	100.21

*THESE COMPONENTS WERE NOT RESOLVED ON THE GAS CHROMATOGRAMS.

out that problems were encountered in two areas.

Because of the large heat effect it was not possible in some instances to operate the reactor isothermally. In the most extreme case the temperature difference between top and bottom of the reactor was of the order of 90°F. Thus the reported temperatures at the milder operating conditions must be considered nominal values only. (The reported temperature is the numerical average of the reactor top and bottom temperatures.) In nonadiabatic-nonisothermal reactors it is not uncommon that the temperature at some point within the reactor will exceed either the top or bottom temperature. Such a phenomenon would go undetected in our experimental set-up since thermocouples were only located at the bed inlet and exit. The actual upper and lower temperatures are presented in Tables 1 and 3.

The second problem was encountered in the analytical portion of the investigation. Three peaks on the chromatogram were found to be mixtures of two components. The unresolved pairs were: 1. Octahydrophenanthrene isomer and n-Butylnaphthalene, 2. Perhydrophenanthrene isomer and n-Butyltetralin and 3. Octahydrophenanthrene isomer and Dihydrophenanthrene. The unresolved "component" yield known to consist of Octahydrophenanthrene isomer and n-Butyltetralin is plotted as a function of temperature (pressure, space velocity constant) in Fig. 3. The curve exhibits two maxima.

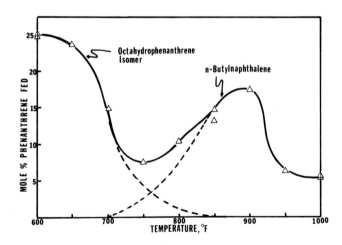

Fig. 3. Yields of Unresolved Octahydrophenanthrene - n-Butylnaphthalene Product Peak.

Thermodynamics considerations suggest that the high temperature maximum is due principally to n-Butylnaphthalene;
whereas, the low temperature maximum is due principally to
Octahydrophenanthrene. The mass spectra of this unresolved
peak from products of the 750°F run and the 800°F run are
consistent with this contention. The dotted line in Fig. 3
is an estimate of the magnitude of each individual contribution to the unresolved peak drawn in such a manner that the
sum of the individual component estimates is equal to the
total. Similar estimates were made for the other unresolved
pairs (6). Separation of the unresolved components in this
manner is admittedly speculative and qualitative. Neverthe-
less this procedure does provide a simplification which is
consistent with the observed data and aids greatly in the
qualitative interpretation of the data.

Since rapid deactivation of the catalyst was expected,
especially at the more severe operating conditions, it was
necessary to maintain a record of declining catalyst activity.
This was done by repeating the selected base conditions, of
850°F, 2000 psi, and 2.0 gm/hr/gm in the first set of experiments and 600°F, 2000 psi, 2.0 gm/hr/gm in the second. While
no significant deactivation was observed in the second set of
experiments (WH-09), Fig. 4 shows that both the conversion of

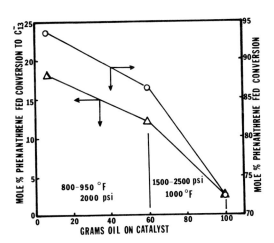

Fig. 4. Conversions at Base Conditions (2000 psig,
850°F, 2 gm/hr/gm).

phenanthrene and the conversion to C_{13}^- decreased with respect
to grams of oil on catalyst in the first set of experiments

(WH-08). As expected, the sharpest decline in catalyst
activity was observed when the more severe operating conditions (1000°F, 1500-2500 psi) were examined.

Yields of the various hydrogenation products of phenanthrene are presented in Tables 2 and 4 and in Fig. 5. Large
quantities of octahydrophenanthrene and perhydrophenanthrene

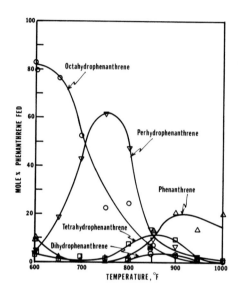

Fig. 5. Yields from the Hydrogenation of Phenanthrene
at 2000 psig, 2.0 gm/hr/gm.

isomers were observed in many of the products. (The various
isomers are lumped together in the figures). At 2000 psig
and a space velocity of 2 gm/hr/gm, octahydrophenanthrenes
are produced in 82% yield at 600°F. With increasing temperatures the octahydrophenanthrenes are further hydrogenated
to perhydrophenanthrenes until a maximum yield of approximately 62% perhydrophenanthrenes is reached at 750°F.
Beyond this temperature the yield of perhydrophenanthrenes
 decreases as the thermodynamic equilibrium is shifted to
favor the less saturated species. Cracking reactions are
also a factor at the elevated temperatures as illustrated
in Figs. 6, 7, and 8. The presence of n-butyldecalin and
decalin in the products indicates that some cracking of
perhydrophenanthrene is taking place. However it appears
that at the temperatures required to hydrocrack perhydrophenanthrene (at 2 gm/hr/gm and 2000 psig) the equilibrium is

shifted away from perhydrophenanthrene formation. No evidence
of large branched paraffins that might be anticipated from

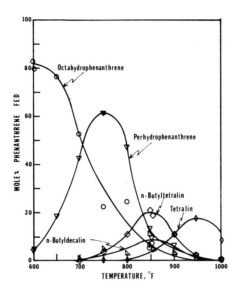

*Fig. 6. Yields from the Hydrogenation of Phenanthrene
at 2000 psig, 2.0 gm/hr/gm.*

mechanisms involving ring opening of perhydrophenanthrene
were uncovered in the mass spectral analyses. Large quantities of tetralins and napthalenes were observed in the

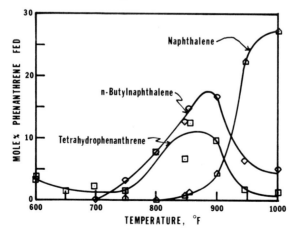

*Fig. 7. Yields from the Hydrogenation of Phenanthrene
at 2000 psig, 2.0 gm/hr/gm.*

cracked products. The presence of large quantities of
n-butane and n-butyl substituted tetralin, naphthalene and
decalin indicates that the major reaction paths involve

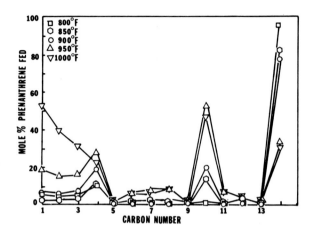

Fig. 8   Product Yields by Carbon Number

saturation and cleavage of terminal rings. In Fig. 9 various
grouped product yields at 950°F and 2000 psig are plotted
versus space time. Again it is evident that the formation
of two ring compounds precedes the formation of one ring
compounds.

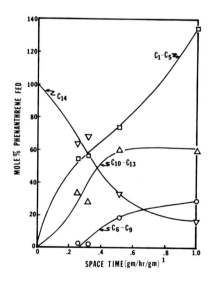

Fig. 9.   Product Yields at 2000 psig, 950°F.

Dealkylation of alkyl substituted aromatics was much more prominent that reactions involving cracking within the side chain suggesting that a carbonium ion mechanism might be operative. In contrast, the product distribution from the thermal hydrogenolysis of phenanthrene indicated that the α-ring-opening of 1,2,3,4-tetrahydrophenanthrene was followed by cracking within the side chain (5). A further piece of evidence suggestive of carbonium ion activity is the indication that some saturated 6-member ring species may have been isomerized to the methyl substituted 5-member ring isomer. The mass spectrum identified as "Octahydrophenanthrene Isomer" in Fig. 2 exhibits a large peak at $m/e=171$ corresponding to the loss of a methyl group during fragmentation.

The only evidence of cracking at the central ring was the presence of trace quantities (less than 1 mole % yield) of biphenyl and cyclohexylbenzene in some of the reaction products. Slightly higher yields of 2-ethylbiphenyl were obtained. It should be pointed out that these data do not entirely dispel the possibility of cracking at the central ring. It would be interesting to conduct some experimentation in the same temperature range but at much lower pressures than employed in the present investigation where the equilibrium yields of dihydrophenanthrene would be higher. Of course catalyst deactivation would likely be an even more serious problem than was encountered in the present study.

Because of the near isothermal behavior of the high temperature runs it is possible to perform a crude analysis of the reaction kinetics. A simple model which appears to describe our system is:

$$A \; \overset{k_1}{\underset{k_2}{\rightleftarrows}} \; B \; \overset{k_3}{\rightarrow} \; C$$

where A represents phenanthrene, B represents hydrogenated product and C represents cracked products (C_{13}^-). If it is assumed that the equilibrium reaction is rapid in comparison with the cracking reaction, then one can show that the overall conversion of phenanthrene to cracked products should follow first order kinetics (6).   But before a meaningful analysis of the data can be undertaken it is necessary that the rate constants be corrected for the observed decline in catalyst activity.   This was done by defining the activity as the ratio of the observed first order rate constant to the first order rate constant obtained by extrapolating the deactivation plot, Fig. 4, to zero grams of oil on catalyst. The activity was then calculated for each yield period using the deactivation curve of Fig. 4. A plot of ln(1-x) versus a(activity) x τ(space time) confirmed that the hydrocracking reaction is indeed correlated with first order kinetics. The activation energy was calculated to be 40 kcal/gmole. Comparison of this value with an activation energy of 65 kcal/gmole estimated from Penninger and Slotboom's thermal data (5) indicates that some catalysis of the cracking reactions is operative.

ACKNOWLEDGEMENTS

The financial support for this investigation was provided by the National Science Foundation under grant AER72-03562 AO2 (formerly GI-36597). The assistance of Dr. Stephen Billets and Dr. Norman E. Heimer in the analysis of the GC-Mass Spectral data is gratefully acknowledged. We would also like to express our appreciation to the Nalco Chemical Company for providing the catalyst.

REFERENCES

1. Pauling, L., The Nature of the Chemical Bond, 3rd Ed., p. 199, Cornell Univ. Press, Ithaca, 1960.

2. Aczel, T., Rev. Anal. Chem. 1, 226 (1972).

3. Curran, G.P., Struck, R.T. and Gorin, E., Ind. Eng. Chem., Process Des. Dev. 6, 166 (1967).

4. Gardner, L.E. and Hutchinson, W.M., Ind. Eng. Chem., Process Des. Dev. 3, 28 (1964).

5. Penninger, M.L. and Slotboom, H.W., Erdol and Kohle, Erdgas, Petrochemie 26, 445 (1973).

6. Huang, C.S., Master of Science thesis, University of Mississippi, May, 1976.

7. Early, W.F. II, Master of Science thesis, University of Mississippi, May, 1976.

NEW MATERIALS FOR COAL LIQUEFACTION

Ricardo B. Levy and James A. Cusumano
Catalytica Associates, Inc.

There is need for catalysts for coal liquefaction
with increased activity, longer life and improved selectivity for desired products. General criteria for these
catalysts are examined. A number of compounds are examined in detail and certain types are suggested for
testing.

I. INTRODUCTION

Direct liquefaction is one of the alternatives currently
under development for the production of clean burning fuels
from coal. In the more advanced liquefaction processes, such
as those of Gulf Research and Development, Hydrocarbon
Research, Inc., and the Bureau of Mines (Synthoil), coal is
liquefied in the presence of a cobalt-molybdate catalyst at
moderate temperatures (400°C) and high hydrogen pressures
(2000-4000 psi). Extensive research is currently underway (1)
to improve the performance of these catalysts. Properties of
importance include activity (to reduce reactor size and
pressure) activity maintenance (to increase life and regener-
ability) and selectivity (to minimize hydrogen consumption).
The catalyst performs a number of functions such as cracking
large aromatic molecules present in coal liquids and hydrogenation of certain cracked molecules with concommittant
removal of heteroatoms such as sulfur, nitrogen and oxygen.
Because of the projected need for clean-burning liquid
fuels and the technological complexity of existing liquefaction technology, first generation processes are likely to
be based on existing catalyst technology. However, it is
clear that there is room for substantial improvements in all
aspects of the process, improvements that will require new
catalytic materials and concepts. It is therefore timely to
establish some of the ground rules which will guide the
catalytic scientist and engineer in the choice of new materials
to be tested as catalysts for coal liquefaction. This was the

objective of a study recently conducted for the Electric Power
Research Institute (2) and summarized in this report.

In this summary, the general criteria for the selection of
materials for coal liquefaction catalysis are discussed first.
This serves as a basis for a detailed analysis of a number of
compounds and a preliminary assessment of the types of compounds that should be tested for coal liquefaction.

II. CRITERIA FOR SELECTION OF MATERIALS

A discussion of new materials requires, in effect, a look
at the enormous number of compounds that have been investigated in solid state and inorganic chemistry over the years.
A number of the most important classes of compounds are shown
in Table 1. They are formed by transition and alkaline earth
metals and a small group of nonmetals from the upper right
hand corner of the periodic table of the elements: boron,
carbon, silicon, nitrogen, phosphorous, oxygen, sulfur and
chlorine.

TABLE 1
Representative Classes of Compounds
==

Oxides	*Simple*	Al_2O_3, MoO_3
	Complex	$Mg_2Mo_3O_8$
Sulfides	*Simple*	V_5S_8
	Complex	$Al_{0.5}Mo_2S_4$
Carbides	*Simple*	WC
	Complex	Pt_3SnC
Nitrides	*Simple*	Co_3N_2
	Complex	V_3Zn_2N
Borides	MoB, $Co_{21}Hf_2B_6$	
Phosphides	Co_2P	
Silicides	Mo_3Si	
Alloys and Inter-metallic Compounds	$Ni-Cu$, $ZrPt_3$	
Organometallic	$Co_2(CO)_8$	
Molten Salts	$ZnCl_2$	
Solid Acids	*Zeolites*, *Clays*	
Solid Bases	CaO, $NaNH_2$	

Considering the wealth of compounds represented by these various groups, it is interesting to note that relatively few have been tested for catalytic applications. Furthermore, it is difficult to choose a priori from this large number of materials those which would be applicable for study as liquefaction catalysts. The challenge in the selection sequence for new materials is therefore the identification of those constraints that have to be met by a compound in addition to its catalytic activity. Among the most important constraints for coal liquefaction are thermal and chemical stability.

Thermal stability takes into account the resistance of a given material to volatility, melting, sintering, and general mechanical failure. Temperatures which must be considered are both those for reaction and regeneration. The reaction temperatures for catalytic liquefaction processes are currently of the order of $400^{\circ}C$. The regeneration scheme which is most likely to be applied to these processes is controlled oxidation of the carbonaceous residues which are deposited on the surface of the catalyst. Although this is carried out with careful control of temperature, surface temperatures can frequently exceed $800^{\circ}C$. Continued use and regeneration brings about structural degradation of many materials at these conditions. Chemical stability relates to the chemical behavior of the materials in the environment of coal liquefaction and regeneration. Of primary concern in this respect is the stability of the catalyst in an H_2S/H_2 atmosphere. In coal liquefaction, concentrations of H_2S of 1-5 percent or higher are normal. Other reactive compounds that may affect stability are hydrocarbons, NH_3, H_2O, and O_2. Except for oxygen used in regeneration, the effect of the other reactants is minimal compared to that of H_2S.

In spite of the lack of catalytic information for many of the compounds shown in Table 1, the focus of a general survey such as the present one will be those materials containing cations that have shown catalytic activity. This is the reason for the emphasis on transition metals compounds. The thermal and chemical stability of these materials will now be discussed.

III. THERMAL STABILITY

The only compounds in Table 1 that are clearly excluded from further consideration because of poor thermal stability are the organometallic complexes. For example, $Co_2(CO)_8$ decompose to the metal at temperatures as low as $150^{\circ}C$, unless the CO pressure in the system is greater than 600 psi (3). By contrast, some of the compounds in Table 1 can withstand

temperatures that are among the highest of any material.
For example, TaC melts at about 3980°C (4) and TiB_2 melts at
2980°C (5). In spite of this high thermal stability, the
chemistry of these compounds changes with temperature and the
stable stoichiometry at the synthesis temperature may be quite
different from that at the temperature of operation. This is
illustrated quite dramatically by a compound that has been
considered for hydrogenation and desulfurization, VS_4 (6).
An examination of a simplified phase diagram for this material
reveals that above 300 to 400°C, VS_4 decomposes to sulfur and
the next stable stoichiometry, V_5S_8 (7) (although there are
indications that V_3S_5 may also be formed (8)). It is therefore
unlikely that at operating conditions VS_4 is the actual
catalyst.

A diagram of temperature vs. composition is not complete
without a specification of pressure. In the case of the V-S
system the pertinent parameter in the pressure of sulfur or,
equivalently, a sulfur containing compound such as H_2S. The
behavior of compounds in the presence of H_2S depends on the
chemical stability of the compound and is discussed in the
next section.

IV. CHEMICAL STABILITY

The two conditions that are most critical in coal liquefaction are the high H_2S concentration and the need (unless
alternate methods are discovered) to use oxygen to regenerate
the spent catalyst. Before these two conditions are discussed,
it is of interest to explore the behavior of compounds in the
presence of the "parent" nonmetallic element, namely the free
energy of formation. For convenience, all the comparisons are
made at 700K (which is comparable to current coal liquefaction
temperature). Thermodynamic calculations are based on the
latest published data and have been discussed in detail
elsewhere (2). An extremely useful simplification, proposed
by Searcy (9), allows direct use of heats of formation in
the absence of values for the entropy change. This permits
considerable extension of the published data and leads to
some interesting conclusions concerning potential new materials
for coal liquefaction catalysis.

A. Free Energy of Formation

By comparing free energies of formation, it is possible
to make a qualitative evaluation of the relative stability of
various compounds. This in turn can be used to predict the

behavior of these compounds in certain chemical environments.
The free energies of formation of a number of compounds were examined in detail (2). Representative examples are shown in Tables 2 and 3. The following general observations

TABLE 2
Standard Free Energies of Formation ($- \Delta G_5^{O}$) of Representative Oxides, Sulfides, Carbides, and Nitrides (In kcal/g-atom Non-Metal at 700 K)

	Oxides		Sulfides		Carbides		Nitrides	
Group IV	TiO	108	TiS	64	TIC	44	TiN	65
	TiO_2	97	TiS_2	39				
Group V	NbO	84	NbS_2	47	NbC	34	NbN	42
	NbO_2	79						
Group VI	MoO_2	55	MoS_2	33	MoC	3	Mo_2N	2
	MoO_3	45						
Group VIII	CoO	42	Co_9S_8	25	Co_2C	-4	Co_3N (unstable)	
			CoS_2	19				

TABLE 3
Standard Free Energies of Formation ($- \Delta G_5^{O}$) of Representative Borides, Silicides, and Phosphides (In kcal/g-atom Non-Metal at 700 K)

	Borides		Silicides		Phosphides	
Group IV	TiB	39	$TiSi$	31	a	
	TiB_2	22	$TiSi_2$	16		
Group V	a		$NbSi_2$	16	a	
Group VI	a		$MoSi$	14	a	
Group VIII	a		$CoSi$	19	CoP	29

a. *Data unavailable.*

are of interest for the present study:

1. *Oxides are the most stable compounds of the groups that were examined.* In effect, the following stability trends are observed:

oxides ≫ nitrides > carbides

oxides > sulfides

oxides ≫ borides, silicides, phosphides

One consequence of these trends is that most compounds are expected to be thermodynamically unstable in an oxidizing environment such as encountered in catalyst regeneration.

2. *In general, the stability of a family of compunds decreases with increasing group number in the periodic table.* Group VIII oxides, for example, are the least stable of the transition metal oxides. However, the extent of this decrease is not the same for all groups of compounds. It is most severe for nitrides and carbides, least severe for silicides. In general, the following order of stability change is observed:

nitrides, carbides ≫ oxides > sulfides > silicides

This difference is manifested in the variation in stability of members of one group of compounds (such as oxides) in the presence of the same environment. In H_2S, for example, titanium oxide (Group IV) is stable. Cobalt oxide (Group VIII) on the other hand, is not.

3. *Thermodynamic information on borides is limited to the Group IV elements Ti, Zr and Hf.* For these elements, borides are more stable than silicides. From the similarity in many of the physicochemical properties of borides and silicides, it is expected that this behavior will continue throughout the periodic table. The behavior of silicides in H_2S can therefore be used as a guide to the stability of borides in this environment.

B.   Stability in the Presence of H_2S

The high H_2S concentrations present during coal lique-faction imposes a most severe constraint on the choice of catalytic materials. Levels as high as 1-5% H_2S can be expected. From the thermodynamics of sulfide formation, it is

found that <u>most</u> materials are <u>unlikely to survive</u> in this
environment (2). Thus, at any reasonable process conditions,
metals, alloys, organometallic complexes, carbides, and many
oxides and nitrides can form the respective sulfide.
However, as indicated earlier, the behavior of individual
transition metals depends on their position in the periodic
table. Some representative examples are shown in Table 4:

TABLE 4
Examples of Stability in the Presence of H_2S
==

	Reaction		$\Delta G^O_{700\ K}{}^a$
Oxides:			
Group IVB	TiO_2	TiS	+22
Group VIB	MoO_3	MoS_2	-33
Nitrides:			
Group IVB	TiN	TiS	+22
Group VB	TaN	TaS_2	-7
Group VIB	*Unstable Nitrides*		$\ll 0$
Silicides:			
Group IVB	$TiSi$	TiS	-18
Group VIIB	$MnSi$	MnS	-12
Group VIIIB	$NiSi$	NiS	+12

   a. *Free Energy of Sulfide formation at 700 K in kcal/
g-atom non-metal. Negative free energy indicates a favored
reaction.*

In general, the following is observed:

1. *While* <u>*oxides*</u> *and* <u>*nitrides*</u> *of Group IV are stable in* H_2S,
those of higher groups can form the sulfide.

2. <u>*Conversely*</u>, *while* <u>*borides*</u> *and* <u>*silicides*</u> *of Group IV are
thermodynamically unstable in* H_2S, *those of Group VIII are
expected to survive even in severe* H_2S *environments.*

3. *Many of the metals that are in the region of intermediate
stability (Groups V, VI and VII) are likely to form complex
compounds, such as oxysulfides, in the presence of H_2S.*
Formation of these compounds is a sensitive function of the
H_2S pressure in the system.

It should be emphasized that the above statements are based
on <u>thermodynamic</u> information only. No conclusions can be
drawn concerning the <u>kinetics</u> of the respective transformations. However, the thermodynamic analysis does provide a
guideline for the expected behavior of a system under the most
adverse conditions.

V.   CONCLUSIONS

Before a material is tested for catalytic coal liquefaction, its chances of survival in the liquefaction
environment should be examined. The presence of H_2S poses
the most severe problem. A large number of compounds that
may ordinarily be considered promising candidates sulfide in
this environment. It is therefore fruitless to spend considerable effort in the testing of these materials. Compounds
that are expected to <u>resist</u> sulfidation include a number of
oxides, nitrides, borides and silicides. Among these there are
a number of interesting compositions that have not been tested
for catalytic liquefaction to-date. Examples are $Mg_2Mo_3O_8$,
which has Mo_3 clusters and has been found to exhibit hydrogenation activity intermediate between metals and oxides (10),
the perovskite-like Nowotny nitrides such as Ni_3AlN, and the
borides of the Group VIII metals such as CoB and NiB. Serious
consideration, of course, should also be given to the large
number of sulfides that have been synthesized and characterized over the last few years, (an example is $Al_{0.5}Mo_2S_4$,
which also contains Mo_3 clusters (11)) and to sulfo-compounds
such as oxysulfides which are likely to be formed by many of
the compounds of intermediate stability. Some of these
are being uncovered only recently, including Ta_2S_2C (12)
which is capable of forming intercalation compounds and also
retains the layered structure that is characteristic of a
number of currently used hydrotreating catalyst.

If an alternative to oxidative regeneration is not found,
even some of the sulfur resistant materials mentioned above
will not be viable candidates for catalytic coal liquefaction
unless they exhibit unusual activity maintenance and therefore
require no or infrequent regeneration. It was observed earlier
that a number of the compounds under consideration are likely
to be thermodynamically unstable in an oxidizing environment.
It is therefore important to consider how they will be

resynthesized to the stoichiometry that is catalytically active. Sulfides, and oxysulfides, of course, present no problem. Carbides and even nitrides may be feasible. The use of PH_3 or B_2H_6 to resynthesize borides and phosphides is probably impractical. This further restricts the best candidates for catalytic liquefaction unless more economical reagents or means for resynthesis are developed. It should be mentioned that even among those classes of materials which may endure oxidative regeneraton (e.g. carbides, nitrides, oxides, sulfides, oxysulfides and mixed systems), numerous compounds exist which are of interest for exploration as future gener-ation liquefaction catalysts.

VI. REFERENCES

(1)   "Catalyst Development for Coal Liquefaction", EPRI Contract No. RP-408-1, Amoco Oil Company.

(2)   Boudart, M., Cusumano, J. A., and Levy, R. B., New Catalytic Materials for Coal Liquefaction, Electric Power Research Institute, Report No. RP-415-1, October 30, 1975.

(3)   Toth, L. E., Transition Metal Carbides and Nitrides. Academic Press, New York, 1971.

(4)   Aronsson, R., Lundstrom, T., and Rundquist, S., Borides, Silicides and Phosphides. Wiley, New York, 1965.

(5)   Gleim, W. T., U. S. Patent No. 3,694,352 (1973).

(6)   F. Fransen and S. Westman, Act. Chem. Scand. 17, 2353 (1963).

(7)   J. Tudo, Rev. de Chimie Minerale t2, 53 (1965).

(8)   M. G. Gaudefroy, Compt. Rend. 237, 1705 (1953).

(9)   J. Tudo and G. Tridot, C. R. Acad. Sc. Paris t258, 6437 (1964).

(10) J. Tudo and G. Tridot, Compt. Rend. 257, 3602 (1963).

(11) W. Blitz and A. Kocher, Z. fur Anorg. und Allg. Chemie 241, 324 (1939).

(12) R. Schollhorn and W. Schmucher, Z. Naturf. 306, 975 (1975).

DEACTIVATION AND ATTRITION OF CO-MO CATALYST DURING H-COAL® OPERATIONS

Cecilia C. Kang and Edwin S. Johanson
Hydrocarbon Research, Inc.
(A Subsidiary of Dynalectron Corporation)

Critical examination of available analytical data on fresh and used catalysts has disclosed that three common causes for catalyst deactivation, namely, sintering, metal deposition, and carbon deposition, contributed to the deactivation of the Co-Mo catalyst used in the H-Coal process. Their relative detrimental effects on the catalyst vary with the rank of coal being processed and the process conditions. The Co-Mo catalyst is supported on gamma alumina. Operations with subbituminous coal such as Wyodak coal caused much more pronounced catalyst sintering than those using bituminous coal such as Illinois No. 6 coal. The high oxygen and high moisture content of Wyodak coal (in the event of incomplete drying) gives rise to a relatively high H_2O/H_2 ratio in the process gas. The pronounced effect of steam on sintering of catalyst at temperatures below 1000°F is well known. Major metal contaminants found on the used catalyst are titanium, iron, calcium, and sodium. The magnitude of titanium deposition on catalyst from bituminous coal feed increased with catalyst age. Titanium deposition was concentrated to a depth of approximately 10% of the radius of spent 1/16" extrudate, whereas iron was found in clusters on the extrudate surface. The used catalyst contained 10 to 35% carbon. The extent of carbon deposition is affected more by process conditions than by either coal feed or by catalyst age. Low pressure operations at 1000 psig or lower yielded high carbon deposition of over 22%. Commercial Co-Mo catalysts have shown different magnitudes of catalyst attrition in the ebullating bed reactor. The current H-Coal catalyst possesses satisfactory attrition resistance as shown by recent PDU operations at gas and coal rates at the design values for the Pilot Plant under construction at Catlettsburg, Kentucky.

The H-Coal® Process has been under development for over ten years and has been supported by government and industry. Work has been carried out on bench-scale units handling from 25 to 100 pounds of coal per day and in a Process Development Unit (PDU) handling 3 tons per day. Some fourteen coals including bituminous, subbituminous, lignite and brown coal have been tested without difficulty.

In the H-Coal Process, coal is dried, pulverized, and slurried with coal-derived oil for charging to the coal hydrogenation unit. The heart of the process is the unique reactor design shown in Fig. 1.

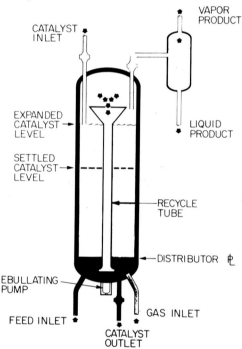

Fig. 1.   Ebullated Bed Reactor

The coal-oil slurry is charged continuously with hydrogen to a reactor containing a bed of ebullated catalyst wherein the coal is catalytically hydrogenated and converted to liquid and gaseous products. In the ebullated bed the upward passage of the solid, liquid, and gaseous materials maintains the catalyst in a fluidized state. The relative size of the catalyst and coal is such that only the unconverted coal, ash, liquid and gaseous products leave the reactor while the catalyst remains therein. Catalyst can be added and withdrawn continuously so a constant activity can be maintained. The reactor provides a simple means of controlling reactor temperature and an effective contact between the reacting species and the catalyst, permitting a satisfactory degree of reaction at reasonable operating pressure. The liquid product from the reactor is a synthetic crude oil which can be converted to gasoline and furnace oil by conventional refining processes. Alternately, under milder operating conditions, a clean fuel gas and low sulfur fuel oils may be produced. The relative amounts of these products depend on the desired sulfur level in the heavy fuel oil.

In September 1974, Hydrocarbon Research, Inc. received an 8.1 million dollar first phase contract, which has been expanded by several million dollars, to design a nominal 600 ton per day Pilot Plant to demonstrate the H-Coal Process. The contract is jointly funded, two thirds by Government and one third by industry. At the present time, two oil companies and Electric Power Research Institute are the industry representatives. Funds are also being provided by the Commonwealth of Kentucky.

The proposed Pilot Plant will be located at Catlettsburg, Kentucky, adjacent to the Ashland Oil Refinery. It is designed for two modes of operation, (a) to process 633 tons of coal per day to produce 1,920 barrels of fuel oil with less than 0.7 W % sulfur from coal containing more than 3 W % sulfur (fuel oil mode), or (b) to process 210 tons of coal per day to produce 740 barrels per day of synthetic crude (syncrude mode). The design basis for the fuel oil mode of operation is based on bench-scale experimentation. The current laboratory program has been directed towards confirming in the PDU the design basis for the Pilot Plant.

I. H-COAL CATALYST

The H-Coal Process uses a commercial Co-Mo catalyst developed originally for desulfurizing petroleum residua. The Co-Mo catalyst selectively hydrocracks the carbon-sulfur bonds to release the sulfur as hydrogen sulfide and to terminate the carbon bond by hydrogenolysis without cracking of the carbon-carbon bond. Bituminous coal and subbituminous coal are characterized by the fact that they are mainly composed of condensed aromatic rings and oxygen-containing functional groups: $-COOH$, $-OCH_3$, $-OH$, and " $C=O$". These first three functional groups are all terminal groups, but the so called "carbonyl group" (" $C=O$") exists as a bridge linking the condensed aromatic rings together. Efficient coal conversion with minimum hydrogen consumption can be achieved if the " $C=O$" linkage is selectively hydrocracked. Since the organic functional groups containing oxygen or sulfur behave similarly, the use of Co-Mo catalyst for coal conversion and desulfurization has given reasonably good performance.

HRI has accumulated over 50,000 hours of process evaluation and optimization in bench scale and process development
units. While the H-Coal Process development is in an advanced
stage in most respects, catalyst evaluation and development
has not kept pace with overall program progress.

II. CATALYST DEACTIVATION

Catalyst deactivation curves in terms of hydrogen consumption and sulfur content of fuel oil are presented in Figs.
2 and 3.

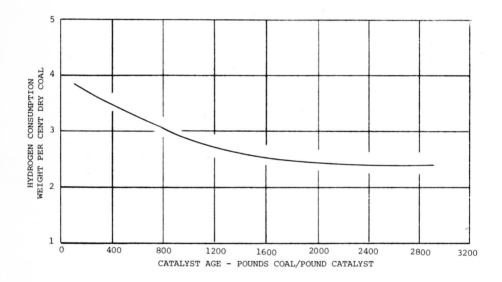

Fig. 2.   Decline in Hydrogenation Activity of Catalyst

These curves were derived from HRI's recent PDU operation with
Illinois No. 6 coal from the Burning Star Mine and were included in a paper presented at the 68th AICHE Meeting, Nov-
vember, 1975 (1).   The deactivation curves are characterized
by a steep initial deactivation followed by a gradual decline.
The shape of the curve is typical of most catalyst deactivation curves.

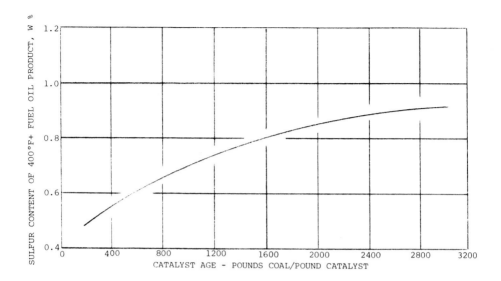

Fig. 3. Decline in Desulfurization Activity of Catalyst

Critical examination of available analytical data on fresh and used catalysts has disclosed that three common causes for catalyst deactivation, namely, carbon deposition, sintering and metal deposition contributed to the deactivation of the Co-Mo catalyst. Their relative detrimental effects on the catalyst vary with the rank of coal being processed and the processing conditions.

A discussion is given below with respect to the magnitude of carbon deposition, sintering, and metal deposition on the Co-Mo catalyst and the relative importance of these three factors in contributing to catalyst deactivation.

A. Carbon Deposition

The used catalysts contained 10 to 35% carbon. In operations in the normal range of conditions the used catalysts contained 10 to 20% carbon. No consistent trend has been observed between the magnitude of carbon desposition on catalyst and the rank of coal feed, although the lower-rank coals (e.g., Wyodak Coal) and lignites tend toward the lower end of the range, while higher rank coals (e.g., Pittsburgh Seam Coal) tend toward the higher end of this range. For purposes of

illustration, Table 1 summarizes the carbon content of used
catalyst from a number of bench unit experiments on Pittsburgh
Seam Coal. The extent of carbon deposition reaches a certain
level within a few days' time and does not increase with further
operations. Low pressure operations at 800 psi hydrogen pressure or lower invariably yielded carbon deposition of over 22%
on used catalyst. Experiments carried out with catalyst after
regeneration indicate that carbon deposition contributes significantly to initial catalyst deactivation.

TABLE 1
Carbon Content of Used Catalysts

Coal Feed	----------Pittsburgh Seam----------				
Coal Processed, Lbs.	760	102	805	2246	2602
Catalyst Age, Hours	503	59	463	1447	263
Hydrogen Partial Pressure, psig	800	-----------1800-----------			
Used Catalyst, % C	26.5	19.0	13.6	20.7	20.0

B. Sintering

The Co-Mo catalysts used in the majority of experiments
are supported on gamma alumina. Gamma alumina is thermally
unstable, particularly in the presence of steam. Literature
data are available on the effect of steam on sintering of alumina. For the purpose of illustration, the effect of temperature and steam on the sintering of a nickel on gamma alumina
catalyst (2) is shown in Fig. 4. The presence of steam not
only causes an initial drop in surface area during the first
sixty hours but also enhances the long term slow rate of decline.

An examination of the available analytical data on fresh,
used, and regenerated catalysts indicates that catalyst sintering occurred after an extended run of thirty days using
Illinois No. 6 coal wherein the $H_2O:H_2$ in the reactor outlet
gas is about 1:8. An eleven-day operation under similar conditions did not cause any noticeable sintering of catalyst.
Sintering was quite pronounced after a forty nine day operation
using partially dried Wyodak coal containing 20% moisture, resulting in a $H_2O:H_2$ ratio of 1:2 in reactor outlet gas. Catalyst sintering could be the cause for the observed high deactivation rate of the wet Wyodak coal operation in comparison
to a similar experiment using more completely dried Wyodak
coal containing only 2% moisture.

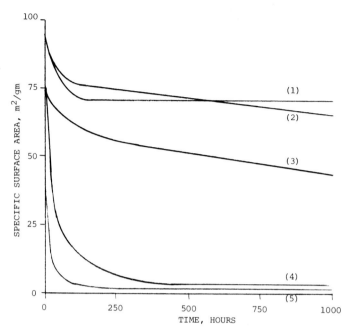

Fig. 4.   Effect of Temperature and Steam on Sintering of Gamma Alumina Catalyst.   (1) 800°C, only H_2, (2) 400°F, $H_2O:H_2=9:1$, (3) 500°C, $H_2O:H_2=9:1$, (4) 600°C, $H_2O:H_2=9:1$, (5) 800°C, $H_2O:H_2=9:1$.

C.   Metal Deposition

     Some typical analyses of coal ash composition and major catalyst contaminants are shown in Tables 2 and 3.   Table 4 presents a semi-quantitative analysis of a typical used catalyst by emission spectrometry.   The relative level of metal contaminants on the catalyst was not in the ratio characteristic of the ash composition of the coal.   Metal contaminants act as a physical barrier between the reactants and active sites of the catalyst.   When any metal contaminant increases with catalyst age, it usually is the prime factor in causing catalyst deactivation.   The common metal contaminants on the used catalyst are titanium, iron, calcium and sodium.   Bituminous coal, such as Pittsburgh Seam and Illinois No. 6, deposits appreciable amounts of titanium and low amounts of calcium and sodium; subbituminous coal, such as Wyodak, usually deposits higher amounts of sodium and calcium than the bituminous coal.   The range of iron deposition is similar with various ranks of coal.

TABLE 2
Composition of Coal Ash

Coal	Wyodak	Illinois No. 6
Ash, W %	9.73	9.95
Composition of Ash, W %		
Phosphorus Pentoxide	0.61	0.06
Silica	33.64	45.74
Ferric Oxide	3.80	16.18
Alumina	17.08	18.95
Titania	1.16	0.95
Calcium Oxide	18.50	9.50
Magnesium Oxide	4.70	0.94
Sulfur Trioxide	18.04	4.52
Potassium Oxide	0.71	1.79
Sodium Oxide	0.80	0.43
Undetermined	0.95	0.94

TABLE 3
Contaminants on Used Catalysts

Coal Feed	Wyodak	Illinois No. 6
Catalyst Age, Lbs Coal/Lb Catalyst	781	2644
Oil-Free Catalyst, Analyses, W %		
Carbon	13.67	10.9
Sulfur	5.13	4.6
Iron	1.73	1.5
Titanium	0.03	5.11
Sodium	1.46	0.42
Calcium	1.04	0.24

TABLE 4
Emission Spectrometry Analysis of a Typical Used Catalyst[a]
==

Element	Approximate W % Metal
Aluminum	10
Antimony	0.02
Arsenic	0.2
Barium	Not detected
Beryllium	0.01
Boron	0.7
Calcium	0.3
Chromium	0.05
Cobalt	1
Copper	0.01
Gallium	0.05
Germanium	0.3
Iron	1
Lead	0.01
Magnesium	0.2
Manganese	0.08
Molybdenum	5
Niobium	0.01
Nickel	0.3
Silicon	0.2
Titanium	3
Vanadium	0.2
Zinc	0.1
Zirconium	0.05

a. Analysis of oil-free (benzene-extracted) + 20 mesh extrudates.

It is observed that among the metal contaminants identified so far, titanium deposition from the bituminous coal increased with catalyst age. Recent analyses of benzene-insoluble and benzene-soluble components of the residua derived from Illinois No. 6 coal show that the benzene-insoluble portion of the residuum contains about the same amount of titanium as present in the M.A.F. coal, 900 ppm, and the benzene-soluble portion of the residuum contains only about 5% of the titanium as present in the coal. Thus, it is quite logical to deduce from these data that titanium exists in a form of organometallic complex in the bituminous coal. This deduction is further supported by the knowledge that titanium is just below vanadium with respect to the strength of the ligand metal bonds and stable vanadium porphyrin exists in petroleum. It should be noted that very little titanium was deposited on the catalyst when Wyodak coal was processed, even though Wyodak coal contains about the same amount of titanium as Illinois No. 6 coal. Further work is needed to establish whether the benzene-insoluble residuum from the Wyodak coal contains any significant amount of titanium. It is also observed that only one-twentieth of the titanium present in the Illinois No. 6 bituminous coal deposited on the catalyst.

Scanning electron micrograph of spent catalysts showed that titanium penetrated to a depth of approximately 10% of the radius of the 1/16" extrudate, and iron was found in clusters on the surface of the extrudate. The way that titanium deposition concentrates in the periphery of the catalyst extrudate is expected to be more detrimental in reducing the effective diffusion of "coal molecule" into the catalyst pore than if the titanium were distributed uniformly throughout the whole extrudate.

When bituminous coal, such as Illinois No. 6, was processed, metal deposition appeared to cause the gradual but steady deactivation of catalyst.

II. CATALYST ATTRITION

The magnitude of catalyst attrition in the ebullated bed reactor is well defined by the rotating drum attrition test developed by HRI. By applying this test, HRI has successfully chosen a commercial catalyst which showed satisfactory attrition resistance during PDU operations. Catalysts recoveries are summarized in Table 5. About 93 W % catalyst (by Mo balance) was recovered after 814 hours' operation, processing 2,687 lbs. coal/lb. catalyst. The increase in weight of the catalyst was consistently greater than the total weight of the major contaminants determined. These and other data obtained with different coal feeds indicate the possibility of the presence of other major contaminants. Deposition of alumina appears to be another contaminant as shown by a few analyses comparing the Al to Mo ratio of fresh and used catalyst.

TABLE 5
Catalyst Recovery from PDU Operations
==

Run	A	B	C	D
Catalyst Age				
Hours	814	535	475	272
Coal Processed,				
Lbs/Lb Catalyst	2687	3061	2451	1509
Catalyst Recovery, W %				
Gross Weight	139.0	140.6	133.8	156.4
Gross Weight less major				
contaminants determined	103.5	102.3	100.0	112.2
Molybdenum Balance	92.7	96.8	96.4	97.6

Recently, a PDU operation was carried out to test for catalyst attrition, catalyst breakdown, and catalyst carryover under fluid dynamic conditions at the high gas velocity intended for commercial operation. The catalyst recovered from this operation was 97.6 W %, basis molybdenum balance. Representative samples of fresh and recovered catalysts were obtained by thieving and rifling. The length of one-hundred particles was measured. The data are presented in Fig. 5

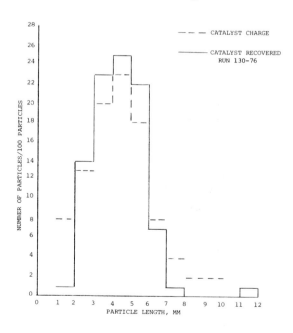

Fig. 5.   Catalyst Particle Size Distribution

The measurements indicate a slight breakdown of particles longer than 6 mm to shorter particles, and a loss of catalyst particles shorter than 2 mm, about 2.4 W % of the total. The results of this run showed that catalyst breakdown is negligible in comparison to the normal catalyst replacement rate, which is defined by catalyst deactivation.

III. SUMMARY AND CONCLUSIONS

Critical examination of available analytical data on fresh and used catalyst disclosed that three common causes for catalyst deactivation, namely, carbon deposition, sintering, and metal deposition, contributed to the deactivation of the Co-Mo catalyst used in the H-Coal Process. Their relative detrimental effects on the catalyst vary with the rank of coal being processed and the process conditions.

The used catalyst contained 10 to 35% carbon. The extent of carbon deposition is affected more by process conditions than by either coal feed or catalyst age. Low pressure operations at 1000 psig or lower yielded a high carbon deposition of over 22%.

The Co-Mo catalyst is supported on gamma alumina. The pronounced effect of steam on sintering of gamma alumina at temperatures below 1000°F is well known. Operations with subbituminous coal such as Wyodak coal caused much more pronounced catalyst sintering than those using bituminous coal such as Illinois No. 6 coal. The high oxygen and high moisture content of Wyodak coal (in the event of incomplete drying) gave rise to a relatively high $H_2O:H_2$ ratio in the process gas, which caused rapid catalyst deactivation through sintering.

Major metal contaminants found on the catalyst are titanium, iron, calcium, and sodium, and in some cases also aluminum. The magnitude of titanium deposition on catalyst from bituminous coal feed increased with catalyst age. Titanium deposition was concentrated to a depth of approximately 10% of the radius of spent 1/16" extrudate, whereas iron was found in clusters on the extrudate surface.

It is concluded that the initial catalyst deactivation is caused by carbon deposition in the case of bituminous coal. Both carbon deposition and sintering contribute to the initial catalyst deactivation with subbituminous coal feed. Titanium deposition from the bituminous coal causes a gradual decline in catalyst activity.

Commercial Co-Mo catalysts have shown different magnitudes of catalyst attrition in the ebullating bed reactor. HRI developed a rotating drum attrition test which defines the attrition resistance of the catalyst under H-Coal processing conditions. HRI successfully chose a commercial catalyst which showed satisfactory attrition resistance during recent extended PDU operations at gas and coal rates at the design values for the Pilot Plant being built at the present time.

IV.   ACKNOWLEDGEMENT

The authors wish to acknowledge the financial support given to this program by ERDA, EPRI, the two oil company sponsors, and the Commonwealth of Kentucky.

V.   REFERENCES

1. Johnson, C.A., Chervenak, M.C., Comolli, A.G., Johanson, E.S., Kang, C.C., and Volk, W., "H-Coal Process Development", 68th AICHE Meeting, November, 1975.

2. Williams, A., Butler, G.A., and Hammonds, J., "Sintering of Nickel - Alumina Catalysts", J. Catalysis 24, 352, (1972).

CATALYTIC LIQUEFACTION OF COAL

Yuan C. Fu and Rand F. Batchelder
Pittsburgh Energy Research Center
U. S. Energy Research and Development Administration

High sulfur bituminous coal is liquefied and desulfurized by hydrotreating with syngas at 3,000 psi in
the presence of added water, vehicle, and catalyst.
Cobalt molybdate catalyst impregnated with alkali metal
compounds, such as potassium carbonate, sodium carbonate, and potassium acetate exhibited good activities
for liquefaction and desulfurization. High coal conversions and oil yields can be obtained in the temperature
range of 400° to 450° C, and the asphaltene and the
sulfur contents of the oil products are comparable to
that obtained in coal liquefaction using pure hydrogen
and cobalt molybdate under similar conditions. Further
improvements in the oil quality could be attained under
more severe conditions at 450°C and increased reaction
time, but both syngas usage and hydrogen usage would
increase substantially. Catalytic coal liquefaction
using syngas gives an improved thermal efficiency and
reduces the capital and operating costs by eliminating
shift converters and purifying systems need for the liquefaction process using hydrogen.

I. INTRODUCTION

Most catalytic liquefaction processes for producing low-
sulfur liquid fuel from coal use large amounts of hydrogen
which will have to be produced at high cost. In our previous
work (1), we reported our attempt to use low-cost syngas to
hydrotreat coal in the presence of added water, vehicle, and
cobalt molybdate-sodium carbonate catalyst. Catalytic coal
liquefaction using syngas reduces the capital and operating
costs by eliminating shift converters and purifying systems
needed for the liquefaction process using hydrogen. New

catalysts have been prepared and tested with syngas to promote liquefaction and desulfurization as well as water-gas shift conversion. Cobalt molybdate catalyst impregnated with alkali metal compounds, such as potassium carbonate, sodium carbonate, and potassium acetate, exhibited good activities for these reactions.

The liquefaction of coal by syngas, like that by hydrogen, appears to proceed via production of asphaltene and conversion of the asphaltene to oil. Because of the important effect of asphaltenes on the viscosity of the product oil (2), the progress of asphaltene formation and asphaltene conversion during the coal liquefaction has been investigated, and some observations on the chemistry of asphaltenes are presented here.

II. EXPERIMENTAL

The liquefaction of coal was studied in a magentically-stirred autoclave. The experimental method is similar to that described earlier (1). The analyses of coal and vehicle used are shown in Table 1. Most coal liquefaction experiments used a liquid fraction (boiling point 270°-630°C) derived from the Solvent Refined Coal Process as the vehicle. For coal hydrotreating by hydrogen alone, the catalyst was a cobalt molybdate supported on alumina-silica (Harshaw CoMo 0402T, 3% CoO - 15% MoO_3). For coal hydrotreating by syngas, the cobalt molybdate catalyst was impregnated with alkali metal compounds, such as potassium carbonate, sodium carbonate, and potassium acetate. The cobalt molybdate-alkali metal compound weight ratio was 2:1. Typically, 2.3 to 2.8 moles of reactant gas and 30 grams of coal as received were charged to the autoclave. The reaction was carried out in the temperature range of 400° to 450° C at about 3,000 psi in the presence of added water, vehicle, and catalyst. Total products were filtered at ambient or warmer temperature to obtain liquid oils. Asphaltenes, operationally defined as being soluble in benzene and insoluble in pentane, were isolated from the oil products according to procedures established by the Analytical Section of the Pittsburgh Energy Research Center. Gaseous products were analyzed by mass spectrometry. Data on conversion, oil yield, and asphaltene formed are given as weight percent based on moisture- and ash-free (maf) coal.

TABLE 1
Analyses of Coal and Vehicle, Percent
===

| | Coal | | | | Vehicle | |
| | Kentucky Bituminous | | W. Virginia Bituminous | | Alkyl- | SRC |
	As used	maf	As used	maf	naphthalene	Liquid
C	58.8	75.9	73.0	80.9	90.6	88.8
H	4.9	5.4	5.3	5.7	8.7	7.4
N	1.2	1.6	1.3	1.4	0.06	1.1
S	5.16	6.67	3.80	4.21	0.37	0.65
O	14.4	10.5	8.6	7.9	0.27	2.1
Ash	15.5		8.1			0.03
Moisture	7.1		1.7			
VM	35.7	46.1	43.3	48.0		
Btu/lb	11,500		12,900			
Asphaltene					0.8	7.1
H/C ratio		0.85		0.84	1.15	1.00

III.   RESULTS AND DISCUSSION

A.   Hydrotreating of Coal

    New catalysts were prepared and tested with syngas for
coal liquefaction. Catalysts were cobalt molybdate impregnated with alkali metal compounds such as potassium carbonate,
sodium carbonate, and potassium acetate. Table 2 shows the
results of hydrotreating West Virginia bituminous coal at
450°C by $2H_2$:1CO syngas. The amount of water added was 10
parts per hundred parts of coal plus vehicle, corresponding
to 0.7 mole of steam per mole of CO in the reactor. The
results obtained with various catalysts were quite
satisfactory.
    Figure 1 shows the increase of syngas or H_2 consumption
with increasing reaction temperature and time. CoMo impregnated with K_2CO_3 was used as the catalyst in the syngas runs.
Regardless of the reaction temperature and time, the
asphaltene content of the oil product decreases with the
increase of the syngas or H_2 consumption as shown in Figure
2. In general, the sulfur content and the oil viscosity also
decrease with the increase of the syngas or H_2 consumption.
Thus further improvements in the oil quality by reduction in
asphaltene and sulfur contents could be attained at the

expense of oil yield under more severe conditions at 450°C
and increased reaction time as shown in Table 3. However,
both syngas and H_2 usage would increase substantially. It
would not be a major problem if low cost syngas were used in
the liquefaction process, but it could be too costly if H_2
were used. It is noteworthy that one objective of the
SYNTHOIL process (3) is to convert coal to a low-sulfur liquid
fuel with minimum consumption of high cost H_2.

TABLE 2
Hydrotreating of West Virginia Bituminous Coal[a]
(Coal:SRC liquid = 1:2.3, 450°C, 15 min.)

Feed gas	Syngas ($2H_2$:1CO)				H_2
Catalyst	K_2CO_3[b]	CoMo-K_2CO_3[c]	CoMo-Na_2CO_3[c]	CoMo-KOAc[c]	CoMo[d]
Water added, parts/100 parts coal + vehicle	10	10	10	10	
Operating pressure, psi	3,100	3,000	3,000	2,800	2,800
Conversion, %	93	95	94	96	94
Oil yield, %	59	68	64	70	71
Asphaltene formed, %	62.9	42.2	52.8	55.5	30.3
S in oil product, %	0.67	0.40	0.45	0.50	0.39
Kinematic viscosity, cst at 60°C	44	20	22	22	15
Syngas or H_2 consumed, scf/lb maf coal	5.4	8.6	10.5	11.5	10.7

a.   Data are given in weight percent of maf coal.
b.   One part per hundred parts coal plus vehicle.
c.   Three parts per hundred parts coal plus vehicle.
d.   Two parts per hundred parts coal plus vehicle.

Table 4 shows the hydrotreating of coal at various temperatures at which the consumption of syngas and H_2 would be
4,000 to 5,500 scf per barrell of oil, the normal range for
a process. The asphaltene and the sulfur contents of the
oil products obtained in the syngas and H_2 runs are
comparable. Under these conditions, the syngas usage was
in the range of 3,900 to 4,700 scf per barrel of oil as
compared to the H_2 usage of 4,700 to 5,500 scf per barrel
when pure H_2 was used. Use of $1H_2$:1CO syngas in place of
$2H_2$:1CO syngas gave no significant difference in the results
except that the H_2/CO ratio of off-gas and CO/H_2 consumption
ratio varied.

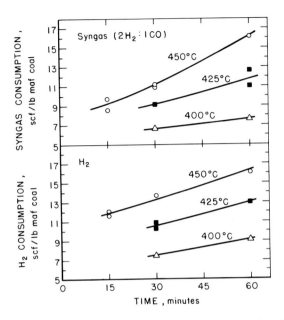

Fig. 1. Change of syngas or H_2 consumption with temperature and time.

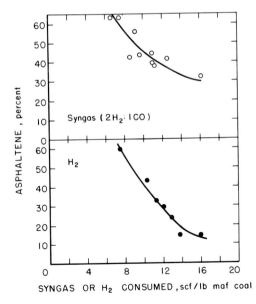

Fig. 2. Decrease of asphaltene with syngas or H_2 consumption.

TABLE 3
Effect of Reaction Time on Hydrotreating of Coal[a]
(Coal:SRC liquid = 1:2.3, 450°C)

Catalyst	Syngas (H_2:CO = 2:1) CoMo–K_2CO_3[b]		H_2 CoMo[c]	
Reaction time, min	30	60	30	60
Water added, parts/100 parts coal + vehicle	10	10		
Operating pressure, psi	3,000	3,000	2,700	2,800
Conversion, %	94	95	95	95
Oil yield, %	65	42	59	48
Asphaltene formed, %	38.2	32.2	14.8	14.7
S in oil product, %	0.43	0.32	0.31	0.26
Kinematic viscosity, cst at 60°C	16	9	10	7
Syngas or H_2 consumed, scf/lb maf coal	11.2	16.2	13.7	16.1
CO/H_2 consumption ratio	1.4	0.8		
H_2/CO ratio of off-gas	4.3	3.2		

a. Data are given in weight percent of maf coal.
b. Three parts per hundred parts coal plus vehicle.
c. Two parts per hundred parts coal plus vehicle.

B. Formation of Asphaltene

The coal liquefaction by syngas, like that by H_2, proceeds via production of asphaltene. The decrease in the asphaltene formation with the increase in the syngas consumption (Figure 2) indicates that the asphaltene formed is converted to oil under further hydrotreating. Sternberg et al. (2) reported a significant effect of asphaltenes on the viscosity of the SYNTHOIL product oil. We also found that the viscosity of the oil product correlates well with its asphaltene content, regardless of the reaction temperature and time (see Figure 3). A similar relationship is also observed when pure H_2 is used. The difference in the curves obtained from syngas and H_2 runs is probably attributable to the difference in the molecular structure and size of asphaltenes obtained (see the following).

TABLE 4

Hydrotreating of Coal at Various Temperatures [a]

(Coal:SRC liquid = 1:2.3)

Catalyst	Syngas (H$_2$:CO = 1:1) CoMo-K$_2$CO$_3$ [b]			Syngas (H$_2$:CO = 2:1) CoMo-K$_2$CO$_3$ [b]			H$_2$ CoMo [c]		
Temperature, °C	400	425	450	400	425	450	400	425	450
Time, min	60	30	15	60	30	15	60	30	15
Water added, parts/100 parts coal + vehicle	10	10	10	10	10	10			
Operating pressure, psi	3,100	3,100	3,100	3,000	3,100	3,000	2,900	2,900	2,800
Conversion, %	89	95	93	89	94	95	89	95	94
Oil yield, %	73	65	64	71	70	68	71	71	71
Asphaltene formed, %	57.2	51.4	53.2	48.9	53.7	42.2	54.4	43.7	30.3
Oil analysis, %									
C	88.2			88.1		88.5	88.3		88.2
H	7.5			7.5		7.5	7.8		7.6
N	1.2			1.2		1.2	1.0		1.1
S	0.54	0.47	0.47	0.53	0.51	0.40	0.40	0.41	0.39
O	2.6			2.7		2.4	2.5		2.7
Kinematic viscosity, cst at 60°C	45	25	21	38	28	20	31	23	15
Syngas or H$_2$ consumed, scf/lb maf coal	7.8	9.9	10.3	7.5	9.2	8.6	9.1	10.3	10.7
CO/H$_2$ consumption ratio	15	23	6.1	3.2	2.2	2.1			
H$_2$/CO ratio of off-gas	1.8	2.1	2.1	3.4	4.4	4.1			

a. Data are given in weight percent of maf coal.

b. Three parts per hundred parts coal plus vehicle.

c. Two parts per hundred parts coal plus vehicle.

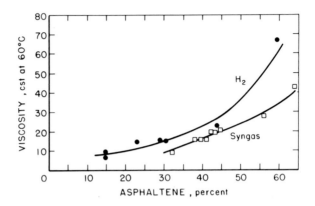

Fig. 3. Oil viscosity vs asphaltene content

Because of the important effect of the amount and the type of asphaltene on the property of the liquefied product, studies were conducted on the formation of asphaltene and oil during the progress of coal liquefaction. Kentucky bituminous coal was liquefied in the presence of an asphaltene-free alkylnaphthalene-based vehicle oil in both syngas and H_2 systems. Catalysts were CoMo - Na_2CO_3 and CoMo for syngas and H_2 runs, respectively. An operating pressure of about 3,000 psi and reaction temperatures of 400° to 450°C were used. It required about 60 to 70 minutes for the autoclave to reach desired temperatures. The moment the desired temperature was reached was taken as zero time.

Figure 4 shows the results at 400°C. Nearly 30% of the conversion occurred before the autoclave reached the temperature in both syngas and H_2 runs. As viewed by earlier workers (4), the rate of coal conversion to asphaltene is far greater than the rate of asphaltene hydrogenation to oil, and the asphaltene formation goes through a maximum with respect to time. The hydrogenation of asphaltene to oil at 400°C is very slow and substantial parts of the conversion product - equivalent to about one-third of maf coal - still remained as asphaltene after 120 minutes. At 450°C, the asphaltene formation appears to reach a maximum rapidly but decreases relatively slowly by hydrotreating (see Table 5). The rates of asphaltene formation and asphaltene conversion are relatively greater in the coal liquefaction by H_2 than by syngas.

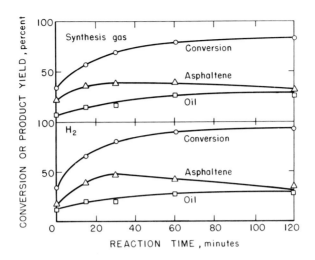

Fig. 4.   Conversion and product yield vs reaction time
at 400°C.

TABLE 5
Formation of Asphaltene During Coal Liquefaction
(Kentucky bituminous coal:alkylnaphthalene = 1:2.3, 2,900
 psi, 450°C)

	Syngas ($2H_2$:1CO)			H_2		
Reaction time, min.	5	15	60	5	15	60
Asphaltene formed, %[a]	42.6	38.0	25.5	47.1	26.7	6.7
Asphaltene analysis, %						
C	84.1	85.8	87.0	85.4	87.1	86.9
H	6.3	6.3	5.0	6.5	6.3	5.7
N	1.6	2.1	2.0	1.5	2.0	2.1
S	1.3	0.15	0.4	0.64	0.23	0.36
O	7.0	5.7[b]	4.0	6.0	4.3[b]	5.1
Molecular weight of asphaltene	387	430	407	457	573	597

a.   Weight percent of maf coal.
b.   By difference.

It is interesting to note that, under similar liquefaction conditions, molecular weight (number average) of the
asphaltene formed under syngas is lower than that of the
asphaltene formed under H_2 (see Table 5).   The molecular
weight was determined by vapor pressure osmometry of the

asphaltene in benzene. Table 6 shows molecular weights of
asphaltene and total oil products obtained from the liquefaction of West Virginia bituminous coal. Except where
specified a cryoscopic method was used. Again, the
molecular weight of asphaltene molecules obtained in the
syngas system is smaller, and this could, at a given
asphaltene content, result in lower oil viscosity in the
syngas system than in the H_2 system (see Figure 3).

TABLE 6
Molecular Weights of Oil Products and Asphaltenes
(West Virginia bituminous coal:SRC liquid = 1:2.3, 450°C,
  3,000 psi)

| Time, min. | Syngas ($2H_2$:1CO) | | H_2 | |
| | Molecular Weight[a] | | Molecular Weight[a] | |
	Asphaltene	Oil product	Asphaltene	Oil product
15	339[b]	218	437[b]	227
30	494	221	544	210
60	395	194	405	194

  a. By cryoscopic method
  b. By osmometry.

C. Liquefaction Process Using Syngas

      In evaluating a coal liquefaction process using syngas,
a modification of the SYNTHOIL process was considered. A
schematic flow diagram is shown in Figure 5, and the gas
stream flows in scfh (on 100 lb/hr maf coal basis) and volume
percent are shown in Table 7. It is assumed that the process
is operated at 3,000 psi and 450°C and that the syngas flow
rate is at a feed gas/coal-oil slurry volume ratio of 5.9.
These operating conditions will result in turbulent flow in a
packed-bed reactor. Since syngas has a higher density than
H_2, a lower volumetric flow is necessary than that of H_2 in
the SYNTHOIL process. The syngas leaving a gasifier with the
equilibrium composition of 1800°F (H_2/CO = 0.53) is introduced to the recycle gas (H_2/CO = 3.4) to make up the feed
gas (H_2/CO = 2). Steam is introduced to the feed gas to
give an H_2O/CO mole ratio of 0.7, corresponding to the autoclave conditions. The flow and composition of the off-gas
(Stream 3) were estimated from vapor-liquid equilibria of
various components with the liquid oil at the receivers.

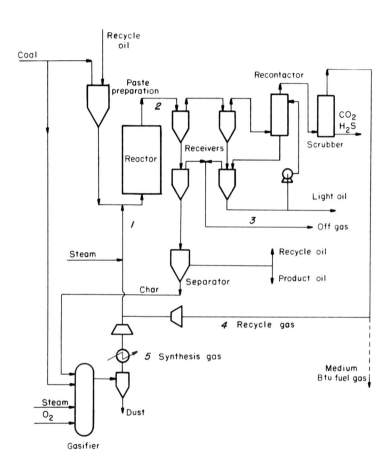

Fig. 5.   Flow diagram of conceptual coal liquefaction
process using syngas.

A small amount of the recycle gas can be removed, if necessary, to keep the methane level from building up. The syngas consumption is about 4,500 scf per barrel of oil when the
oil yield is 3.5 barrels per ton of coal (as received).
Since the syngas produced from the gasifier is used directly,
shift converters and purifying systems required in the
SYNTHOIL Process can be eliminated.

TABLE 7
Gas Stream Flows
(Basis: 100 lb/hr maf coal feed to process)
==

	Gas composition, mole percent				
Stream	1	2	3	4	5
H_2	46.4	47.6	15.8	61.7	30.6
CO	22.7	14.7	13.5	18.1	58.3
CH_4	12.2	15.4	20.7	18.4	0
C_2H_6	1.2	1.9	6.9	1.8	0
H_2S	0.2	0.5	1.8	0	1.3
CO_2	1.6	12.9	41.3	0	8.7
H_2O	15.7	7.0	0	0	1.1
Flow, scfh	4774.7	4206.4	336.7	3162.9	873.6

The thermal balance of the process is shown in Figure 6.
The calculation was based on 100 Btu input of coal. Heating
values are: coal as received = 12,900 Btu/lb, oil = 17,000
Btu/lb, recycle gas = 475 Btu/scf, and off-gas = 439 Btu/scf.
The overall thermal efficiency was calculated to be 76.6%.

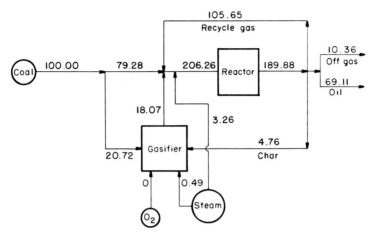

$$\text{Thermal efficiency} = \frac{69.11 + 10.36}{100 + 0.49 + 3.26} = 76.6\%$$

Fig. 6. Thermal balance of liquefaction process (basis: 100 Btu input of coal).

IV. CONCLUSIONS

Liquefaction of high sulfur bituminous coal at 3,000 psi under syngas in the presence of steam, recycle vehicle, and cobalt molybdate catalyst impregnated with potassium carbonate gives high coal conversions and oil yields at 400° to 450°C. The asphaltene, the sulfur content, and the viscosity of the oil products decrease with increasing consumption of syngas, and the syngas consumption increases with increasing reaction temperature and residence time. When syngas is used in place of hydrogen in coal liquefaction, the rate of asphaltene conversion to oil is slower but asphaltenes formed have molecules of smaller sizes.

A catalytic coal liquefaction process using syngas gives high thermal efficiency and reduces the capital and operating costs by eliminating shift converters and purifying systems.

V. ACKNOWLEDGEMENT

The authors thank Steven Macaruso and John Ozanich for their assistance.

VI. REFERENCES

1. Fu, Y. C., and Illig, E. G., Ind. Eng. Chem., Proc. Design Develop., 15, 392 (1976).
2. Sternberg, H. W., Raymond, R., and Akhtar, S., ACS Symp. Series No. 20, 111 (1975).
3. Akhtar, S., Mazzocco, N. J., Weintraub, M., and Yavorsky, P. M., Energy Communications 1, 21 (1975).
4. Weller, S., Pelipetz, M. G., and Friedman, S., Ind. Eng. Chem., 43, 1572 (1951); ibid., p. 1575.

Reference to a company is made to facilitate understanding and does not imply endorsement by the U. S. Energy Research and Development Administration.

KINETICS OF COAL HYDRODESULFURIZATION IN A BATCH REACTOR

R.C. Koltz, R.M. Baldwin, R.L. Bain, J.O. Golden, J.H. Gary
Colorado School of Mines

ABSTRACT

The kinetics of the hydrodesulfurization of coal was studied at the Colorado School of Mines under ERDA contract. The purpose of the study was to investigate the rate of removal of total sulfur from a bituminous coal during batch hydrogenation.

Experimentation was done in a 300 cc Magnedrive reactor. Reaction temperatures of 360, 390, and 420°C were studied at times of 1/2, 1, 2, and 4 hours. Fixed operating variables were: reaction pressure, solvent type, coal type, and solvent-to-coal ratio. The effects of heat up time were minimized by heating the bulk of the solvent to reaction temperature and then injecting a small sample of coal-solvent slurry. Total sulfur analysis was performed on the feed material, solid product, and liquid product; the gaseous products were analyzed with a gas chromatograph.

The results obtained indicated that the desulfurization of the coal increased with an increase in either reaction temperature or reaction time. A reaction rate constant was determined which appeared to be dependent upon conversion as well as temperature. At low conversion levels the reaction rate constant showed a true Arrhenius temperature dependence.

I. BACKGROUND

In recent years there has been a renewed interest in developing this nation's coal reserves because of our growing reliance on foreign oil sources. Due to the stringent air quality standards set by the Environmental Protection Agency, however, the burning of coals with sulfur contents of greater than one percent has been essentially prohibited.

An alternative to the production of low-sulfur coal reserves exists in the desulfurization of high-sulfur coal reserves by the techniques of solvent refining and hydrogena-

tion. It is generally accepted (1) that sulfur occurs in coal
in three forms: organic, pyritic, and sulfate. These forms
vary in concentration and ease of removal from one coal to
another. The purpose of this paper is to develop a kinetic
model that will represent the rate of removal of sulfur for
all three forms simultaneously. Since several different sulfur reactions occur at the same time it is feasible to consider a model in which the reaction rate appears to vary as a
function of conversion. Such models were successfully applied
by Hill, et al. (2) to the dissolution of coal in tetralin
and by Lessley, et al. (3) to thermal cracking of shale gas
oil under a hydrogen atmosphere.

Much of the earlier work done on coal desulfurization
took the form of carbonization studies in which coal or coal
char was heated in the presence of various gas streams and the
percent removal of total sulfur from the coal was determined.
A good review of the work done on this subject prior to 1932
is given by Snow (4).

In 1960 Batchelor, et al. (5) published an article describing a method in which a bed of char was fluidized with a
known mixture of hydrogen and hydrogen sulfide to establish
the equilibrium distribution of sulfur between gas and char.
He also developed an equation for calculating the maximum
amount of desulfurization that could be achieved.

In more recent years a non-isothermal method for determining the kinetics of coal desulfurization has been developed
in which the sample is subjected to a constant rate of heat.
Vestal, et al. (6) suggests that this method is superior to
isothermal methods since it avoids the uncontrolled occurrence
of chemical reactions during the time that the sample is being
heated to reaction temperature. A good review of this method,
complete with theory, experimental procedure and apparatus,
results, and discussion, is given by Yergey, et al. (1). The
study reveals that in most cases the kinetics of hydrogen sulfide removal can be described by five processes. These processes are directly related to the five forms of sulfur present in the coal which are designated as Organic I, Organic II,
Pyrite, Sulfide, and Organic III. With each of the processes
there is an associated activation energy, reaction order, and
rate constant.

Another kinetic model for desulfurization is given by
Qader, et al. (7). This article discusses the hydroremoval of
sulfur from coal tars and concludes that the reaction is first
order with respect to heterocyclic molecules. The experimental results also show that sulfur removal follows a true Arrhenius temperature dependence.

To date no articles dealing with the kinetics of hydrodesulfurization of coal in liquid phase have appeared, but
several articles have been published on coal dissolution

kinetics.

Hill (8) proposed a model for coal dissolution in which a series of reactions take place between the solvent and the coal residue.   The model is given as follows:

$$\text{Solvent} + \text{Coal} \xrightarrow{K'_0} R_0 + L_0 + G_0$$

$$\text{Solvent} + R_0 \xrightarrow{K'_1} R_1 + L_1 + G_1$$

$$\text{Solvent} + R_1 \xrightarrow{K'_2} R_2 + L_2 + G_2$$

$$\vdots$$

$$\text{Solvent} + R_{N-1} \xrightarrow{K'_N} R_N + L_N + G_N$$

where
   R_i is the solid coal residue,
   L_i is the extract in solution,
   G_i is the gaseous product.

In another article Hill, et al. (2) develop a model in which the first-order reaction velocity constant varies with the fraction of coal extracted. This model fits the kinetic data in the range of $350°$ to $450°C$ quite well. Plots of the Arrhenius energy of activation and the Eyring enthalpy of activation are included in the article and both plots exhibit straight line relationships.

Wen, et al. (9) have proposed a rate equation for the dissolution of coal under hydrogen pressure which describes fairly closely the experimental data reported from two independent sources. Wen's equation describes the rate of dissolution as a function of the fraction of undissolved solid organics and the coal-solvent ratio. It also incorporates an Arrhenius temperature dependence and an exponential dependence on the hydrogen partial pressure.

Another semi-empirical correlation which adequately represents coal dissolution data is discussed by Curran, et al. (10). This correlation deals more specifically with the mechanism of hydrogen transfer and does not lend itself to the application of conventional kinetic data analysis techniques.

II.   EXPERIMENTAL DESIGN

The purpose of this study was to determine the effects of temperature and time on the desulfurization of coal and to develop a kinetic model that can satisfactorily represent the rate of total sulfur removal.

The coal used in this study was a bituminous coal from the Madisonville No. 9 seam, Fies Mine, in Kentucky. The proximate and ultimate analyses for the coal are given in Table 1. This coal was selected because it is currently being used in the start up of the Fort Lewis ERDA solvent refining plant. Coal of minus 200 mesh size fraction was used for this study.

TABLE 1

Proximate and Ultimate Analyses of Coal Used[a]

(A) Runs 1–11

Proximate Analysis	As Received	Dry Basis
% Moisture	6.00	–
% Ash	16.20	17.20
% Volatile	32.80	34.90
% Fixed Carbon	45.00	47.90
	100.00	100.00

Ultimate Analysis		
% Carbon	62.90	66.90
% Hydrogen	4.60	4.89
% Nitrogen	1.10	1.17
% Sulfur	2.86	3.04
% Oxygen	11.34	6.80
% Ash	17.20	17.20
	100.00	100.00

(B) Runs 12–14

Ultimate Analysis		
% Carbon	64.50	65.70
% Hydrogen	4.41	4.49
% Nitrogen	1.33	1.35
% Sulfur	3.40	3.46
% Oxygen	9.56	7.90
% Ash	16.80	17.10
	100.00	100.00

TABLE 1 (Continued)

(C) Runs 15-17

Ultimate Analysis	As Received	Dry Basis
% Carbon	61.00	63.50
% Hydrogen	3.88	4.04
% Nitrogen	1.34	1.39
% Sulfur	3.66	3.81
% Oxygen	11.60	7.96
% Ash	18.50	19.30
	99.98	100.00

[a]Coal: Fies Mine, Source: Kentucky, Rank: Bituminous.

The solvent used was straight run anthracene oil purchased from the Reilley Tar and Chemical Company. The raw
solvent was vacuum distilled at an absolute pressure of 2-3 mm
of mercury and the cut between $125^{o}-250^{o}C$ was saved. This
cut is similar to the cut used in the Pittsburgh and Midway
Solvent Refined Coal Process (11). Hydrogen gas for this
study was 3500 psig grade with a purity of 99.95%. The gas
was manufactured by the Linde division of the Union Carbide
Corporation.

The temperatures chosen for this study were 360°, 390°,
and $420^{o}C$. The lower temperature was selected since literature (12) indicates that at temperatures below $350^{o}C$ the dissolution of coal in the solvent is incomplete. The upper temperature was selected because at temperatures of greater than
$450^{o}C$ coking occurs.

Studies on the kinetics of dissolution of coal (2) indicate that at times of greater than 4 hours the percent dissolution does not significantly increase. For this reason
reaction times of 1/2, 1, 2, and 4 hours were selected. The
lower limit was selected because the 10 minutes required for
heating up the injected slurry would interfere with runs of
less than 30 minutes.

Initial pressures of 750, 785, and 820 psig were used
for runs at 420°, 390°, and 360° respectively. These initial
pressures resulted in a reaction pressure of approximately
1900 psig.

The solvent-to-coal weight ratio was set at 10 to 1.
This ratio was chosen because it kept the amount of slurry
injected into the reactor at a minimum.

The determination of the percent of sulfur remaining in
the solvent refined coal was done by an ASTM total sulfur
method. This number was then corrected to give the amount of
sulfur remaining in the coal on a solvent free basis. Sulfur

analyses were also made on the reclaimed solvent and the reaction off gas. Based on these analyses and the analysis of the
original coal and solvent a sulfur balance was completed for
each run.

III.   EXPERIMENTAL EQUIPMENT

All experimental runs in this study were carried out in
a 300 cc Magnedrive batch autoclave, manufactured by Autoclave
Engineers of Erie, Pennsylvania. A manual Ruska piston pump
(250 ml capacity) was used for injection of slurry into the
autoclave. An equipment flow sheet is presented in Figure 1.

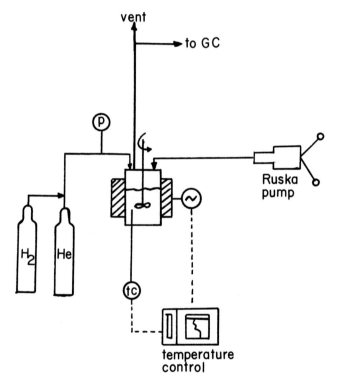

Fig. 1.   Experimental Apparatus.

IV.   EXPERIMENTAL PROCEDURE

One hundred eighty milliliters of vacuum distilled solvent were added to the reactor and the head sealed. The reactor was then purged with helium and pressurized with hydrogen
to the desired initial (cold) pressure, and the jacket heater
turned on. One hundred grams of a thick paste (1:1 ratio by
weight) of 200 mesh coal and solvent was then charged to the

Ruska pump and all air bled from the system. The reactor was allowed to heat (with constant stirring) to reaction temperature, at which time exactly 40 ml of slurry were charged to the hot reactor via the Ruska pump. At the conclusion of the reaction, the reactor was quenched by dropping the heating jacket and cooling the autoclave vessel with a high speed fan. Product gas was analyzed on a gas chromatograph and solvent recovered by vacuum distillation of the resulting liquid product.

V. RUN CONDITIONS

Table 2 shows the run numbers and the corresponding reaction conditions. Runs 1-11 were all performed using the same coal. Different samples of coal were used for runs 12-14 and 15-17 because an insufficient quantity was prepared initially.

TABLE 2
Run Conditions

Run No.	Reaction Temperature (^{o}C)	Reaction Time (min)
1	420	120
2	420	30
3	420	60
4	360	60
5	390	240
6	360	30
7	390	120
8	360	240
9	390	30
10	390	60
11	420	240
12	360	120
13	390	60
14	420	120
15	360	30
16	360	30
17	360	30

VI. PERCENT DESULFURIZATION OF THE COAL

A plot of sulfur conversion for each temperature level as a function of time is given in Figure 2. The data points for 360°C were fit by the method of least squares for a straight line. The data points for 390° and 420° were fit with a flexible curve. Attempts to fit these data points with second and higher order polynomials proved unsuccessful and

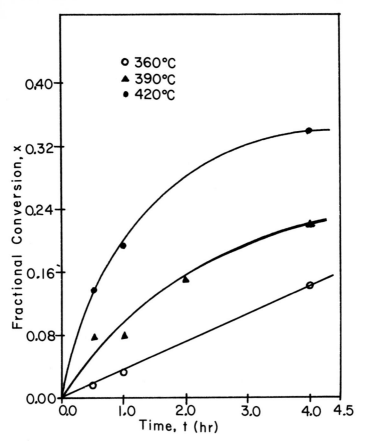

Fig. 2.   Fractional Conversion.

there was no theoretical basis for trying to fit the data with other mathematical models.

VII.   CORRELATION OF THE DATA

Initial attempts to plot the data according to a rate equation of nth-order proved unsuccessful. Although the data at 360°C fit a first-order model quite well the data at the higher temperatures would not yield straight lines for any simple rate expression. The fact that a constant value could not be obtained for the rate constant suggested that the rate constant might be a function of some other variable such as the fractional conversion of the sulfur compounds to hydrogen sulfide and desulfurized products. This would seem feasible since several sulfur reactions are occurring simultaneously (1) with different rate constants and activation energies for each reaction. The idea of the reaction rate varying as a

function of conversion is not a new one and has found application in several areas. Hill, et al. (2) developed a model
which successfully described the rate of dissolution of coal
in tetralin and Lessley, et al. (3) developed a similar model
for the thermal cracking of shale gas oil under a hydrogen
atmosphere. Other applications are described by Fabuss, et al.
(15) to the thermal decomposition rates of saturated cyclic
hydrocarbons and Buekens, et al. (16) to the thermal cracking
of propane.

The model of Hill, et al. (2) proved successful in representing the data taken in this study and is developed below.

Rearranging the rate expression for a simple first-order
irreversible reaction yields the equation

$$\frac{dx/dt}{1-x} = k.$$

For each temperature level the value of dx/dt was evaluated at several different times by using the method of "Equal
Area Graphical Differentiation" as described by Fogler (17).
A plot of $(dx/dt)/(1-x)$ vs x was then made (see Figure 3),
and the data exhibited a linear relationship. The best fit
straight line through each data set was determined by a least
squares fit.

The linear change of the rate constant, k, with x, the
fraction converted can be expressed as

$$k = C_1 - C_2 x$$

$$k = C_1 \left(1 - \frac{C_2}{C_1} x\right)$$

If $C_1 = k_o$ and $C_2/C_1 = a$

then $k = k_o(1-ax)$ (1)

The values of k_o and a were found by rearranging the
coefficients of the best fit straight line to the form of
equation (1).

Substituting equation (1) into the first-order rate expression gives:

$$\frac{dx}{dt} = k_o(1-ax)(1-x)$$

where k_o is a pseudo second-order rate constant.

Separating the variables and integrating gives:

$$\frac{dx}{(1-ax)(1-x)} = k_o dt \quad \text{and}$$

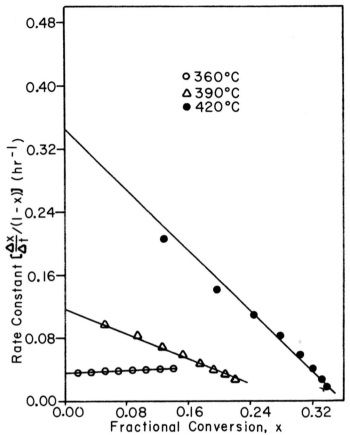

Fig. 3. Rate Constant vs. Fractional Conversion.

$$\ln\left(\frac{1-x}{1-ax}\right) = (k_o t + C)(a-1)$$

Using the boundary condition: t = 0, x = 0, the value of C is found to equal 0. Therefore, the final equation becomes

$$\ln\left(\frac{1-x}{1-ax}\right) = k_o t (a-1)$$

The values of k_o and a for this equation are listed in Table 3 as a function of temperature.

VIII. ARRHENIUS ACTIVATION ENERGY

Based on the values of k_o given in Table 3 a plot of ln k_o vs 1/T was made to determine the Arrhenius energy of activation. This plot is shown in Figure 4. The linear relationship indicates that the desulfurization reactions follow a

TABLE 3
Values for the Parameters k_0 and a
==

Temperature (°C)	k_0	a
360	0.03544	−1.165
390	0.1166	3.373
420	0.3454	2.787

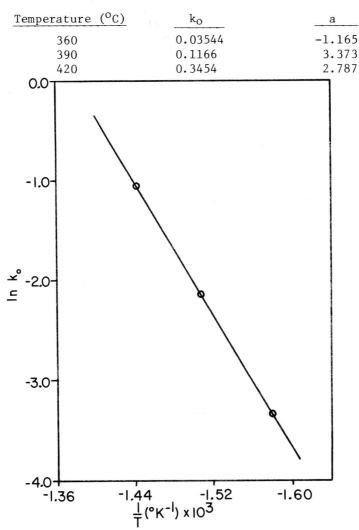

Fig. 4. Arrhenius Plot.

true Arrhenius temperature dependence at low conversions. The
value of the activation energy obtained from this graph is
33.04 Kcal/mol.   This value is within the range of reported
values for hydrodesulfurization reactions (1).

IX. REPRODUCIBILITY

Runs 15, 16, and 17 were carried out under the same set of conditions to serve as a check on the reproducibility of the data. The conditions chosen were a temperature of $360^{\circ}C$ and a reaction time of 1/2 hour. This set of conditions represents an extreme that should give the maximum variance in the results. At the other extreme the large reaction time would tend to minimize the effect of the temperature drop after injecting the slurry. The percentage of desulfurization for these three runs are shown in Table 4 along with the mean and standard deviation.

TABLE 4
Reproducibility Results

Run Number	Percent Desulfurization
15	19.42
16	21.26
17	20.21
Mean	20.30
Standard Deviation	0.92

X. SULFUR BALANCES

A sulfur balance was completed for each run based on the weights and sulfur contents of the reactants and products. In no case was the weight of the sulfur in the products more than 0.4 grams less than the weight of the sulfur in the reactants. It is quite probable that these sulfur losses can be attributed to the volatilization of sulfur-containing compounds during the vacuum distillation. The results of the sulfur balances, expressed as percent recovery are shown in Table 5.

XI. CONCLUSIONS

The following conclusions can be made from this study.
(1) The percent desulfurization of coal is a function of both time and temperature. Increasing either of these variables within the range of conditions for this study will cause the conversion to increase.
(2) The reaction rate constant appears to be a variable of conversion as well as temperature. The relationship between these variables can be adequately described by an equation of the form

$$k = k_o (1-ax)$$

where k_o and a are constants.

TABLE 5
Sulfur Balance Results

Run No.	Percent Recovery
1	91
2	91
3	95
4	97
5	91
6	97
7	93
8	97
9	99
10	93
11	79
12	83
13	83
14	77
15	95
16	93
17	91

The general expression showing conversion as a function of time is then given by

$$\ln\left(\frac{1-x}{1-ax}\right) = k_o t (a-1).$$

This expression represents the kinetic data taken in this study quite well.

(3) At low values of conversion the reaction rate constant shows a true Arrhenius temperature dependence. The value of the activation energy as calculated from the Arrhenius plot is 33.04 Kcal/mole.

(4) The desulfurization of coal is affected to a large extent by the nature of the coal. Even coal taken from the same sample will give large variations in the percent of desulfurization if it is not carefully mixed.

LITERATURE CITED

1. Yergey, A. L., Lampe, F. W., Vestal, M. L., Day, A. G., Fergusson, G. J., Johnston, W. H., Snyderman, J. S., Essenhigh, R. H., and Hudson, J. E., "Nonisothermal Kinetics Studies of the Hydrodesulfurization of Coal," Industrial and Engineering Chemistry Process Design and Development, vol. 13, no. 3 (1974), 233-240.

2. Hill, G. R., Hariri, H., Reed, R. I., and Anderson, L. I.,
"Kinetics and Mechanism of Solution of High Volatile
Coal," Coal Science, Advances in Chemistry Series, vol.
55, Published by the American Chemical Society, Washington, D.C. (1966), 427-447.

3. Lessley, G. E., Silver, H. F., and Jensen, H. B., "Thermal
Cracking of Shale Gas Oil Under a Hydrogen Atmosphere,"
Preprints, Division of Petroleum Chemistry, Inc., American
Chemical Society, vol. 15, no. 4 (1970), A84-A92.

4. Snow, R. D., "Conversion of Coal Sulfur to Volatile Sulfur
Compounds During Carbonization in Streams of Gases,"
Industrial and Engineering Chemistry, vol. 24, no. 8
(1932), 903-909.

5. Batchelor, J. D., Gorin, E., and Zielke, C. W., "Desulfurizing Low Temperature Char," Industrial and Engineering
Chemistry, vol. 52, no. 2 (1960), 161-168.

6. Vestal, M. L., and Johnston, W. H., "Chemistry and Kinetics of the Hydrodesulfurization of Coal," Preprints,
Division of Fuel Chemistry, American Chemical Society,
vol. 14, no. 1 (1970), 1-11.

7. Qader, S. A., Wiser, W. H., and Hill, G. R., "Kinetics of
the Hydroremoval of Sulfur, Oxygen, and Nitrogen from a
Low Temperature Coal Tar," Industrial and Engineering
Chemistry Process Design and Development, vol. 7, no. 3
(1968), 390-397.

8. Hill, G. R., "Experimental Energies and Entropies of
Activation - Their Significance in Reaction Mechanism
and Rate Prediction for Bituminous Coal Dissolution,"
Fuel, vol. 45, no. 4 (1966), 329-340.

9. Wen, C. Y., and Han, K. W., "Kinetics of Coal Liquefaction," Preprints, Division of Fuel Chemistry, American
Chemical Society, vol. 20, no. 1 (1975), 216-233.

10. Curran, G. P., Struck, R. T., and Gorin, E., "Mechanism
of the Hydrogen-Transfer Process to Coal and Coal
Extract," Industrial and Engineering Chemistry Process
Design and Development, vol. 6, no. 2 (1967), 166-173.

11. Anderson, R. P., and Wright, C. H., "Coal Desulfurization
in the P and M Solvent Refining Process," Preprints,
Division of Fuel Chemistry, American Chemical Society,
vol. 20, no. 1 (1975), 5.

12. Guin, J. A., Tarrer, A. R., Taylor, Z. L., and Green,
    S. C., "A Photomicrographic Study of Coal Dissolution,"
    Preprints, Division of Fuel Chemistry, American Chemical
    Society, vol. 20, no. 1 (1975), 66-78.

13. American Society for Testing and Materials, Book of
    A.S.T.M. Standards, Pt. 19, Gaseous Fuels; Coal and
    Coke, D271-68, 1916 Race St., Philadelphia, Pa. (1970),
    23-25.

14. American Society for Testing and Materials, Book of
    A.S.T.M. Standards, Pt. 17, Petroleum Products,
    D1552-68, 1916 Race St., Philadelphia, Pa. (1970), 574-
    580.

15. Fabuss, B. M., Kafesjian, R., Smith, J. O., and Satter-
    field, C. N., Thermal Decomposition Rates of Saturated
    Cyclic Hydrocarbons," Industrial and Engineering
    Chemistry Process Design and Development, vol. 3, no. 3
    (1964), 248-254.

16. Buekens, A. G., and Froment, G. F., "Thermal Cracking of
    Propane," Industrial and Engineering Chemistry Process
    Design and Development, vol. 7, no. 3 (1968), 435-446.

17. Fogler, S., "Equal Area Graphical Differentiation," The
    Elements of Chemical Kinetics and Reactor Calculations,
    Prentice-Hall, Englewood Cliffs, New Jersey (1974),
    468-471.

18. Frolich, P. K., Tauch, E. J., Hogan, J. J., and Peer,
    A. A., "Solubilities of Gases in Liquids at High Pres-
    sure," Industrial and Engineering Chemistry, vol. 23,
    no. 5 (1931), 548-550.

KINETICS AND SOLUBILITY OF HYDROGEN
IN COAL LIQUEFACTION REACTIONS

Dr. James A. Guin, Dr. Arthur R. Tarrer
Wallace S. Pitts, and Dr. John W. Prather
Auburn University

Design Optimization of the SRC process may seek to
minimize hydrogen consumption and maximize the yield of
boiler fuel. Tests on Kentucky No. 9/14 coal were made
to determine hydrogen solubility and solution rates.
Observed rates were very high. These were compared with
rates of hydrogen consumption by reaction. Mass trans-
resistance appears negligible and the consumption rate
is governed by chemical kinetics.

I. INTRODUCTION

Interest in the kinetics of coal dissolution and heteroatom removal in the presence of donor solvents and hydrogen
gas stems from a desire to understand the process and from a
need to obtain data from which coal liquefaction processes
may be designed. If reliable kinetic expressions can be
obtained, the effects of various operating variables quantified, and the trends better established, then data taken in
laboratory experiments can be more confidently extrapolated
to other conditions with a minimum of experimental investigation.

From work in this laboratory (1,2) and that of others (3,
4,5,6) a free radical mechanism for the dissolution of coal in
hydrogen donor solvents can be postulated. The initial dissolution of the coal solid is thermally initiated; however,
the net rate of depolymerization for a given coal depends upon
the nature of the solvent and its effectiveness in stabilizing
the free radicals. The higher the hydrogen donating ability
of the solvent, the more effective the solvent is in terminat-
ing radicals and promoting coal solvation. This is shown by
the fact that hydrogenated recycle solvent has been found to
facilitate coal solvation much more readily than untreated
solvent (18). The overall rate limiting step in the process

appears to be the rehydrogenation of the donor vehicle. This latter process can, however, be aided significantly by the action of coal mineral matter (7,8). The separation of these two steps, hydrogen transfer and solvent rehydrogenation, may provide the key to an improved SRC type process, allowing greater reactor throughput and operation at lower temperatures and pressures. Such an arrangement would be similar to the coal conversion process at Cresap, West Virginia.

Considerable research has been conducted concerning the kinetics of coal solvation and sulfur removal. Although performed in the absence of a solvent, the recent study of the nonisothermal kinetics of coal hydrodesulfurization reported by Yergey, et. al. (9) is of interest. These investigators divided the sulfur in coal into five classes of Organic I, II, III, Pyritic, and Sulfide, and determined pre-exponential and activation energies for the reaction of each of these five types. Of particular interest was the reverse reaction of H_2S with the organic matter of coal to product the Organic III class of coal-contained sulfur. Liebenburg and Potgieter (10), and Potgieter (11), have recently reported studies on the uncatalyzed hydrogenation of coal and on the kinetics of conversion of tetralin during the hydroliquefaction of coal. They concluded that heating and cooling could cause considerable spurious effects in batch autoclave kinetics studies and suggested sampling techniques to avoid these. Kinetic rate constants for the formation of asphalt and oil fractions from the uncatalyzed hydrogenation of coal in tetralin were reported and several different kinetic mechanisms were postulated. Wen and Han (12) have also determined rate constants using coal liquefaction and desulfurization data gathered primarily from studies by Pittsburgh and Midway Coal Mining Co., the University of Utah, and the Colorado School of Mines. These researchers were able to fit data from these sources with an empirical expression for the rate of coal dissolution; however, no kinetic expression was obtained for desulfurization, probably because of lack of sufficient data. Furthermore, the effect of hydrogen partial pressure was not firmly established, and it is likely that the different type reactors and experimental procedures employed in the three laboratories made the data correlation more difficult. Coal liquefaction data using creosote oil together with a CO-stream gas phase have been reported by Handwerk, et. al. (13) at the Colorado School of Mines. In these experiments it was established that reaction temperature had a stronger effect on desulfurization than did hydrogen partial pressure; however, reaction rate expressions were not reported. Similar investigations employing synthesis gas, with the addition of an external catalyst, have been reported by Fu and Illig (14) and Appell, et. al. (15). With regard to catalytic studies, the University of Utah studies on hydrocracking and heteroatom removal (16, 17) are represen-

tative of batch autoclave studies in the presence of an added
catalyst, although in these particular studies a coal-derived
oil was used as a starting material rather than a coal-solvent
slurry. Given et. al. (18) at the Pennsylvania State University, in cooperation with Gulf Research and Development Company, have reported recent results of their efforts to corre-
late coal liquefaction behavior with chemical and physical
characteristics of the coal. Although their results are en-
lightening, it appears that further work will be required to
firmly establish the relation between different characteristics
of the coal and liquefaction behavior, in view of the large
number of interacting variables which are present. Thus, despite the fact that numerous studies have been reported concerning liquefaction kinetics, it appears that the reaction-
dissolution process is still not completely understood and
that further investigation is needed.

   In contrast to the large amount of research on the kinetics of coal solvation and sulfur removal, there is a distinct
lack of related information on the kinetics of hydrogen consumption during the coal solvation process. The rate of hydrogen consumption is important in the SRC process where it is
desirable to minimize the consumption of costly hydrogen and
maximize the yield of the SRC boiler fuel. The production of
highly hydrogenated products including C_1-C_4 gases decreases
SRC yield and increases the consumption of hydrogen. The
synthesis of boiler fuel for use in power plants does not require extensive hydrogenation. It is interesting to note that
SRC product usually contains a slightly lower hydrogen/carbon
ratio (H/C = 0.75) than the feed coal itself (H/C = 0.8) (19,
26). Thus, if the production of light gases (H/C = 6.6),
water, and light liquids (H/C - 1.5) can be minimized, the SRC
process has the potential to be self-sufficient in hydrogen
which can be recovered from light liquids and gases by steam
reforming, if necessary. The SRC pilot plants operated at
Wilsonville, Alabama, and Tacoma, Washington, currently consume
about two weight percent of hydrogen per pound of MAF coal
feed. Controlling solvent composition and process conditions
to optimize selectively the production of SRC product would
limit hydrogen consumption to a minimum and still product an
environmentally acceptable boiler fuel. Before optimum conditions can be selected, however, kinetic rate expressions are
needed to provide models for process scale-up, simulation, and
optimization. This work reports the development of such a
model for hydrogen consumption in the dissolver stage of the
SRC process.

II. REACTION KINETICS EXPERIMENTS

A. Reagents and Materials

Kentucky No. 9/14 coal mixture was crushed; and the -170 mesh fraction - having the screen analysis shown in Table 2, and the elemental analysis in Table 3 - was used. The creosote oil used in this work was furnished by Southern Services, Inc., and was used as received from Wilsonville. Typical analysis of this creosote oil by gas chromatography is given in Table 1. The oil is distilled from coal tar in the boiling range 175° to $350^{\circ}C$. The oil was originally obtained from Allied Chemical Company as creosote oil 24-CB; and has a carbon-to-hydrogen ratio of 1.25 (90.72% C and 6.05% H), a specific gravity of 1.10 at $25^{\circ}C$, and a boiling point range of 350 to $650^{\circ}F$. A 3:1 solvent-to-coal weight ratio was used in all experiments reported, and all coal was dried overnight at $100^{\circ}C$ and 25 inches of Hg vacuum before use. Hydrogen was obtained from Linde Hydrogen in 6000 psi grade and had a purity of 99.995%. Practical grade mesitylene was obtained from Matheson Coleman and Bell (MCB) and used without further purification.

B. Procedures

For each run, a 30 gm. of coal/90 gm of solvent slurry was charged into a 300 cc. magnedrive autoclave from Autoclave Engineers, Inc. Reactions were carried out at reaction times of 15, 30, 60, and 120 minutes and at reaction temperatures of 385, 400, 410, and $435^{\circ}C$; a stirrer setting of 1000 rpm was used in all the runs, with the exception of one run in which a stirrer setting of 2000 rpm was used to evaluate mass transfer effects. A heat-up rate of about $20^{\circ}C$ per minute was used - requiring only about three minutes for heat-up within the zone in which significant reaction occurs (above $370^{\circ}C$) and a total heat-up time of about 30 - 35 minutes. Prior to heat-up 400 psig of hydrogen was charged to the reactor and at reaction temperature more hydrogen was added to give the desired initial hydrogen partial pressure. During each run reaction temperature was held constant within $\pm 3^{\circ}C$; and upon completion, the final hydrogen partial pressure (psia) was determined from gas analysis and total pressure measurement.

TABLE 1
Gas Chromatographic Analysis of Creosote Oil

Compound	Weight %
coumarone	.10
p-/cymene	.02
indan	.11
phenol	.12
o-cresol	.05
benzonitrile	.12
p-cresol	.37
m-cresol	.16
o-ethylaniline	.03
naphthalene	5.1
thianaphthene	.08
quinoline	.37
2-methylnaphthalene	1.3
isoquinoline	.30
1-methylnaphthalene	.38
4-indanol	.55
2-methylquinoline	.42
indole	.21
diphenyl	.49
1,6-dimethylnaphthalene	.39
2,3-dimethylnaphthalene	.19
acenaphthene	6.0
dibenzofuran	6.7
fluorene	10.3
1-naphthonitrile	.18
3-methyldiphenylene oxide	1.7
2-naphthonitrile	.14
9,10-dihydroanthracene	2.4
2-methylfuorene	.85
diphenylene sulfide	.52
phenanthrene	18.6
anthracene	4.3
acridine	.19
3-methylphenanthrene	.98
carbazole	2.2
4,5-methylenephenanthrene	2.5
2-methylanthracene	.24
9-methylanthracene	1.2
2-methylcarbazole	1.7
fluoranthene	5.5
1,2-benzodiphenylene oxide	.96
pyrene	2.6

TABLE 2
Screen Analysis of Bituminous Kentucky No. 9/14 Coal Mixture

Mesh Size of Screen	% Retention
170	1.23
200	1.92
230	1.09
270	4.30
325	17.94
400	10.86
-400	62.65
Total	99.99

TABLE 3
Chemical Analysis of Bituminous Kentucky No. 9/14 Coal Mixture

H	4.9
C	67.8
Total Sulfur	2.55
Organic Sulfur	1.63
FeS_2	0.79
Sulfate Sulfur	0.13
Total Ash	7.16

III.   SOLUBILITY OF HYDROGEN IN COAL LIQUIDS

Data on the solubility of hydrogen in the coal-solvent slurry and a knowledge of the Henry's law constant is necessary in the kinetic modeling to follow hydrogen consumption. In addition, hydrogen solubility data are of importance for design and analysis of subsequent hydrogen recovery and downstream hydrogenation units in the coal processing train.

The coal-derived process solvent for liquefaction operations is typically a complex mixture consisting of partially hydrogenated polynuclear aromatic compounds, capable of transferring hydrogen to the coal structure. The exact chemical composition of the steady-state recycle solvent is dependent upon the characteristics of the feed coal and operational conditions. The creosote oil having the composition given in Table 1 was used as the start-up solvent at the SRC pilot plant in Wilsonville, Alabama, and is reasonably representative of the steady-state recycle solvent. Solubility data for hydrogen in this oil are determined and used for the kinetics modeling in this investigation.

Determinations of the solubility of hydrogen in coal liquids have not been previously reported at coal liquefaction conditions. Peter and Weinert (20) have determined hydrogen solubility in slack wax, a paraffin oil, under similar conditions as encountered here. Eakin and Devaney (21) have measured hydrogen solubilities in paraffinic, naphthenic, and aromatic solvents as a ternary system with hydrogen sulfide in the temperature range from 100° to 400°F and a pressure range from 500 to 2000 psia. Chappelow and Prausnitz (22) have also made measurements of the Henry's law constants for hydrogen solubility in squalane and actamethylcyclotetrasilozane between 25° and 200°C. The work presented here provides data on hydrogen solubility in creosote oil solutions at the high partial pressures of hydrogen and high temperatures used in coal liquefaction reactors. Thus, these data may be applied directly to coal conversion systems.

IV. HYDROGEN SOLUBILITY EXPERIMENTS

The equilibrium cell used was a one gallon Autoclave Engineers 316 Stainless Steel magnedrive autoclave. Equilibrium pressures were monitored with a Heise Bourdon-tube gauge which had been calibrated against a deadweight gauge. The temperature of the cell was measured with a Type K thermocouple inserted into a thermowell extending into the liquid phase. Temperature was controlled within $\pm1\%$.

Gases encountered in this experiment were analyzed for H_2, Air, CH_4, CO_2, H_2S, C_2H_6, C_3H_8, $i-C_4H_{10}$, $n-C_4H_{10}$, $i-C_5H_{12}$, and $n-C_5H_{12}$. Hydrogen analyses were carried out on a Varian Model 920 areograph using a 15 foot column packed with 75% molecular sieve 13X and 25% molecular seive 5A. The column was operated isothermally at 100°C with nitrogen as a carrier gas (40 ml/min). The other gases were analyzed on a Varian model 1800 areograph using a 15 foot column packed with Porapak Q, 80-100 mesh. The column temperature was programmed between 40° and 230°C at approximately 12°C/min. Helium was used as a carrier gas (40 ml/min). Both chromatographs were equipped with thermal conductivity detectors.

The creosote oil was charged into the autoclave and the system purged by evacuation. The autoclave was then brought to the desired temperature and hydrogen added to an amount determined by observation of the pressure. Hydrogen pressure was found to reach an equilibrium state in less than 5 seconds by standard step-response testing. Stirring was carried on during the entire process (2000 rpm), except during sample withdrawal.

Samples of the vapor were withdrawn from the top of the autoclave and analyzed by gas chromatography. From the analysis of the vapor phase and the knowledge of the gauge pressure and the vapor pressure of the creosote oil (Figure 1), the partial pressure of hydrogen in the vapor was calculated.

Liquid samples were withdrawn from the bottom of the autoclave into a stainless steel bomb. The bomb was then fitted to an evacuated glass rack, (Figure 2), in which the volume of the dissolved gases were measured. After volume measurement the gases were passed to a gas chromatograph for analysis. From this analysis - knowing the barometric pressure, ambient temperature, and total gas volume - the weight of hydrogen dissolved in the creosote oil was computed using the ideal gas law. By weighing the loaded bomb the weight of the oil was determined inferentially, and the solubility calculated as grams H_2/gram oil.

V.   RESULT OF HYDROGEN SOLUBILITY EXPERIMENTS

To verify the liquid sampling and analytical procedures the solubility of H_2 in mesitylene was determined at 400°F (204°C) at partial pressures of hydrogen between 500 and 2000 psia. A comparison of the data obtained for solubility of hydrogen in mesitylene in this work to the literature values (21) is given in Figure 3. The measured values agree with the previously reported values within a few per cent, thus confirming the techniques used here.

Experimental data for the solubility of hydrogen in creosote oil were obtained at temperatures of 100°, 200°, 300°, and 400°C at partial pressures of hydrogen ranging from 500 to 3000 psia; and the resulting data are shown in Figure 4. Hydrogen solubility exhibits an interesting inverse temperature behavior, with solubility at 400°C being greater than values at 100°C at the same pressure.

The major sources of error in this work are associated with transfer of liquid sample from the autoclave to the gas burette. The sample bomb was massive and determination of the weight of oil withdrawn was the least accurate step in the experiment. Minor errors result from uncertainties in the temperature of the equilibrium cell and the measurement of dissolved gas volume. As a result of these factors, the present data are estimated from least squares analysis to have an experimental accuracy of 4-6% in solubility at a given partial pressure of hydrogen.

It is apparent from these data that hydrogen is appreciably soluble in creosote oil. Thus in a batch autoclave with equal volumes of gas and liquid at 400°C and 2000 psig

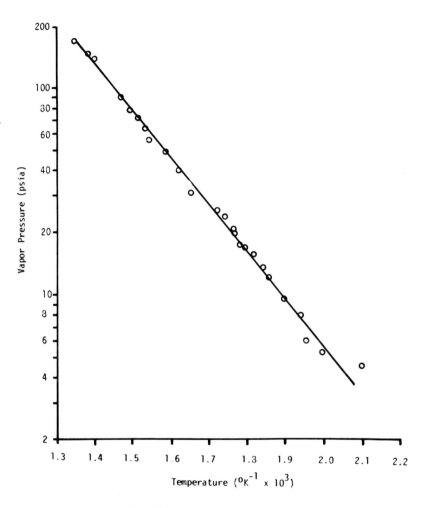

FIGURE 1. Vapor Pressure of Creosote Oil

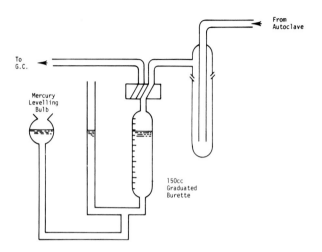

FIGURE 2.   Gas Solubility Measurement Apparatus

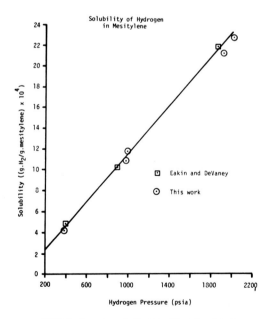

FIGURE 3.   Comparison of this data with previous data for
hydrogen solubility in Mesitylene.

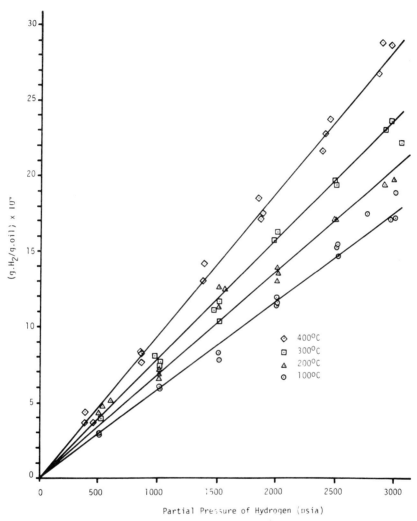

FIGURE 4. Solubility of H_2 in Creosote Oil

approximately one-third of the total hydrogen gas present is
dissolved in the liquid phase.

In order to determine the effect of coal on the hydrogen
solubility, experiments were run in which a slurry of 3:1
ratio of solvent to coal was used instead of creosote oil.
The data were taken at 400°C and 2500 psia total pressure
after 30 minutes at temperature. Assuming all the organic
coal matter to be in the liquid phase, the hydrogen solubility
was essentially the same - that is, within experimental error -
as that in creosote oil alone at the same conditions. Thus,
the hydrogen solubility data in Figure 4 may also be used in
the subsequent study of coal liquefaction kientics.

VI.   CONTROLLING REGIMES IN COAL LIQUIDS HYDROGENATION

The three-phase - solid-coal, hydrogen-gas, and donor-
solvent - reaction system present in the SRC process is subject to several possible mass transport effects. The fact
that coal particles readily disintegrate in the presence of
an appropriate donor solvent (1) and that initial particle
size seemingly has little effect upon the rate of solvation
(23) indicates that pore diffusion and fluid-solid mass
transfer play minor roles in the SRC process - though additional research is desirable to fully substantiate this tentative conclusion. Nonetheless, the rate of gaseous hydrogen consumption might be controlled, at least partly, by mass
transfer across the gas-liquid interface. When gas-liquid
mass transfer controls the overall reaction rate, the most
important factor is the interfacial area, which is governed
by the agitation rate. To test for the presence of mass
transfer regulation, experiments were performed using autoclave stirring speeds of 1000 and 2000 rpm respectively (24).
The experimentally measured gas phase composition indicated
no difference in the net rate of hydrogen consumption, thus
indicating the absence of mass transfer resistance for the
hydrogenation reaction. Further evidence is provided by
Figure 5 where, after injecting a pulse of hydrogen as, a
new quasi-equilibrium state is rapidly established, immediately following hydrogen saturation of the liquid phase. The
data of Figure 5 were obtained by allowing the reaction mixture of equilibrate at 405°C and 700 psig hydrogen pressure.
The pressure was then increased very rapidly to 1870 psig,
by opening and closing the hydrogen inlet valve. The pressure rise and decay was followed by a Leeds and Northropp
pressure transducer. The rapid approach to equilibrium shown
in Figure 5 indicates that at any given time, the gas and
liquid phases are in equilibrium with respect to hydrogen

concentration, unless the reaction is very fast indeed.

A third and final criterion for the absence of mass transfer influence upon the hydrogen consumption rate is the magnitude of the activation energy subsequently determined. The activation energy of 21 kcal/mole determined experimentally is indicative of kinetic control rather than diffusional control. Thus, it appears that mass transfer is not rate controlling in the noncatalytic (except for mineral matter effects) hydrogenation of coal solutions and that lumped parameter, homogeneous-phase reaction rate expressions are adequate for kinetics modeling.

VII. REACTION KINETICS MODELING

The purpose of this section is to determine the magnitude of the parameters in an appropriate kinetic rate expression for hydrogen consumption. Material balances for hydrogen in the gas and liquid phases in the batch autoclave may be written:

$$\frac{dH_g}{dt} = -\dot{m} \tag{1}$$

$$\frac{dH_L}{dt} = -r_A v + \dot{m} \tag{2}$$

The mass transfer term $\dot{m}$ allows for the transfer of hydrogen from the gas to the liquid phase as the reaction proceeds and hydrogen is depleted. The form of the reaction rate r_A must be determined from experimental data. Addition of Equations 1 and 2 eliminates the mass transfer term $\dot{m}$ and yields Equation 3 for the rate of disappearance of total hydrogen:

$$\frac{dH_T}{dt} = \frac{d(H_g + H_L)}{dt} = -r_A V_L \tag{3}$$

For a first order reaction of dissolved hydrogen in the liquid phase, the form of r_A is

$$r_A = k_L H_L / V_L \tag{4}$$

Using the solubility data presented in Figure 4 we may determine the Henry's law constants for use in the relation:

$$\beta = S / P_{H_2} \tag{5}$$

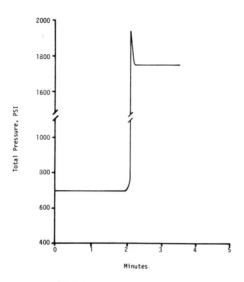

FIGURE 5.   Rate of Approach to Gas + Liquid Equilibrium

TABLE 4
Henry's Law Constants for Hydrogen in Creosote Oil

$T_1 °C$	$\beta \times 10^7$, g H_2/g oil-psia
$100°$	5.95
$200°$	6.94
$300°$	7.75
$400°$	9.65

TABLE 5

$T_1 °C$	α
$385°$	0.183
$400°$	0.193
$410°$	0.196
$435°$	0.211

Representative values of β are given in Table 4. Equation 5 may be used together with the experimental parameters to yield Equation 6.

$$\alpha = H_L/H_T \qquad (6)$$

Values of α for the experimental temperatures used herein are presented in Table 5. Employing the assumption of quasi-equilibrium discussed previously, Equations 3, 4, and 6 may be solved to yield the significant result:

$$\frac{H_L}{H_{LO}} = \frac{H_G}{H_{GO}} = \frac{H_T}{H_{TO}} = \exp(-\alpha\, k_L\, t) \qquad (7)$$

Note that the solubility of hydrogen enters the kinetics model through the parameter α. Equation 7 gives the amount of hydrogen present in either the gas or liquid phase as a function of time during the reaction. A comparison of Equation 7 with the experimental data is presented in Figure 6. The satisfactory fit verifies the assumption of first-order kinetics made in Equation 4. Values of k_L obtained from the slopes are presented in the Arrhenius plot of Figure 7, where an activation energy of 21 kcal/mole for the hydrogenation reaction has been determined. The complete experimental data for this investigation is recorded elsewhere (25). The total amount of hydrogen consumed by the reaction at any time may be found according to

$$H_c = H_{TO}\,\{1-\exp(-\alpha\, k_L\, t)\} \qquad (8)$$

VIII.   CONCLUSIONS

The rate of consumption of hydrogen in coal solvation can be adequately described by a homogeneous kinetic rate expression first-order in dissolved hydrogen concentration. Mass transfer influence appears to be negligible and the overall hydrogen consumption rate is governed by chemical kinetics alone. The reaction rate constant has a frequency factor of 1.06×10^5 per minute and an activation energy of 21 kcal/mole. It is likely that these numerical values are affected by the mineral matter present in the coal, which can catalyze hydrogenation activity (8).

The solubility of hydrogen in coal liquefaction solutions is appreciable and may be represented adequately by a Henry's law coefficient. The solubility exhibits an inverse temperature-solubility behavior.

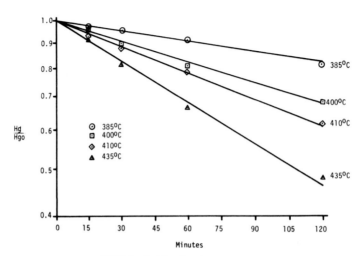

FIGURE 6.   Verification of Hydrogenation Model

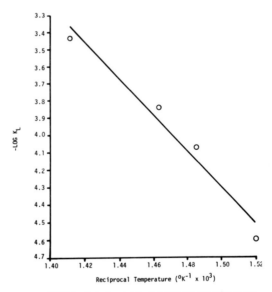

FIGURE 7.   Arrhenius plot for hydrogen transfer rates to coal.

A. Notation

H = mass of hydrogen, g
k = first order rate constant, min^{-1}
$\dot{m}$ = rate of mass transfer from gas to liquid, g/min
P - pressure, psia
S = solubility, g H_2/g oil
r_A = reaction rate of hydrogen in liquid phase, g/min-cc
T = temperature, ^oK
t = time, min
V = volume, cc
α = parameter defined by Equation (5)
β = Henry's law constant, g H_2/g oil-psia

B. Subscripts

c = amount consumed by reaction
g = in gas phase
L = in liquid phase
T = total amount of both places
0 = amount at t = 0

C. Acknowledgements

The authors are grateful to the RANN division of the National Science Foundation for support of this work under Grant No. 38701. They also wish to thank the personnel of Southern Services, Inc., who kindly supplied coal and solvent samples for this work, as well as for their many helpful discussions and suggestions during the course of the work.

IX. REFERENCES

1. Guin, J. A., Tarrer, A. R., Green, S. C., "Mechanistic Study of Coal Dissolution" I & E C Process Design and Develop., (1976) in press.
2. Tarrer, A. R., and Guin, J. A., "Photomicrographic Studies of Coal Particle Dissolution" 68th Annual AIChE Meeting, November 16-20, Los Angeles, Ca.
3. Curran, G. P., Stuck, R. T., and Gorin, E., "The Mechanism of Hydrogen Transfer Process to Coal and Coal Extract", Am. Chem. Soc., Div. of Fuel Chem., Vol. 10, No. 2, C130 (1966).

4. Kloepper, D. L., Rogers, T. F., Wright, C. H., and Bull, W. C., "Solvent Processing of Coal to Produce a De-Ashed Product", Research and Development Report No. 9, prepared for the OCR by Spencer Chemical Division, Gulf Oil Corporation (December, 1965).

5. Oele, A. P., Waterman, H. I., Goedkoop, M. L. and Van-Krevelen, D. S., "Extractive Disintegration of Bituminous Coals", Fuel, 30, 169 (1951).

6. "Homogeneous Catalytic Hydrocracking Processes for Conversion of Coal to Liquid Fuels" ERDA report No. Fe-2202-1 Stanford Research Institute, Menlo Park, CA (Feb., 1976) p. 12.

7. Wright, C. H., and Severson, D. E., "Evidence for the Catalytic Effect of Mineral Matter" ACS Div. of Fuel Chem. Preprints, 16 (2) 68 (1972).

8. Tarrer, A. R., Guin, J. A., Prather, J. W., Pitts, W. S., and Henley, J. P., "Effects of Coal Minerals on Reaction Rates in Coal Liquefaction" ACS Div. of Fuel Chem. Preprints, This Volume, (1976).

9. Yergey, Alfred L., Lampe, F. W., Vestal, M. L., Day, A. G., Fergusson, G. J., Johnston, W. H., Sryderman, J. S., Essenhigh, R. H., Hudson, J. E., Ind. Eng. Chem. Process Des. and Develop. 13 233 (1974).

10. Liebenberg, Barend J. and Potgieter, Hendrik G. J., Fuel 52, 130, 1973.

11. Potgieter, Hendrik G. J., Fuel 52, 134, (1973).

12. Wen, C. Y. and Han, K. W., ACS Division of Fuel Chemistry Preprints, 20 (1), 216, (1975).

13. Handwerk, J. C., Baldwin, R. M., Golden, J. O., and Gary, J. H. ibid., 26.

14. Fu, Y. C. and Illig, E. G., ibid., 47.

15. Appell, H. R., Moroni, E. C., and Miller, R. D., ibid., 58.

16. Qader, S. A., and Hill, G. R., Ind. Eng. Chem. Process Des. and Develop., 8, 450, (1969).

17. Qader, S. A., Wiser, W. H., and Hill, G. R., Ind. Eng. Chem. Process Des. and Develop. 7, 390 (1968).

18. Given, P. H., Cronauer, D. C., Spackman, W., Lovell, H. L., Davis, A. and Biswas, B. Fuel, 54, 40, (1975).

19. "New Catalytic Materials for the Liquefaction of Coal" EPRI Report No. 415, prepared by Catalytica Associates, Inc., Palo Alto, CA (Oct., 1975).

20. Peter, S., and Weinert, M., Z. Physik, Chem. (Frankfort) 5, 114 (1955).

21. Eakin, B. E., and DeVaney, W. E., Thermodynamics - Data and Correlations, AIChE Symposium Series, 70 (140), 30 (1974).

22. Chappelow, C. C., and Prausnitz, J. M., AIChE J., 20, 1097 (1974).

23. Cudmore, J. F., Inst. of Engineers, Australia Mech. and Chem. Eng. Trans. 4 173 (1968).

24. Henley, J. P., "Experimental Evidence of Catalytic Activity of Coal Minerals in Hydrogenation and Desulfurization of Coal Liquids" M.S. Thesis, Auburn University, Auburn, AL (1975).

25. Pitts, W. S., "Reaction Kinetics in the SRC Process", M. S. Thesis, Auburn University, Auburn, Al (1976).

26. Draeger, A. G., "Solvent Refined Coal Pilot Plant: Analysis of Operations," Technical Report No. 6, prepared for Southern Services, Inc., by Catalytic, Inc., under Contract 34210, (Aug. thru Jan., 1975).

SHORT CONTACT TIME COAL LIQUEFACTION
Techniques and Product Distributions(1)

D. D. Whitehurst and T. O. Mitchell
Mobil Research and Development Corporation

ABSTRACT

Initial results are reported on investigations
of coal liquefaction by the solvent refined
coal process. A description is given of the
design and operation of a short contact time
reactor in which reaction times as short as
15 seconds can be achieved. The details of
several runs and early key findings are given.
Coal dissolution is very fast and requires
very little hydrogen consumption. The presence
of H_2 gas in the early stages of conversion is
not critical, but a good H-donor solvent must
be present. Sulfur and oxygen are removed in
a kinetically-parallel fashion. About 40% of
each may be removed readily and rapidly with
little or no H-consumption; thereafter, considerably more hydrogen is consumed than the
stoichiometry requires for the production of
H_2S and H_2O. The initial products of coal
dissolution contain significant amounts of high
molecular weight material which is rapidly
converted to low molecular weight products.
The highest SRC yield is obtained early in the
reaction process; improvement of SRC quality
is accompanied by a decrease in yield and a
large increase in hydrogen consumption.

---

(1)   This work was carried out under a jointly-
      sponsored contract between Mobil Research and
      Development Corporation and the Electric Power
      Research Institute, Contract #RP-410-L

I.   INTRODUCTION

In our investigation of the nature and origin
of asphaltenes in processed coals we have concentrated on the chemistry and kinetics of reactions
accompanying the dissolution of coals. These investigations have been facilitated by the development of two new techniques for the study of coal
liquefaction reactions and their products. First,
we have designed and operated a 300-ml batch autoclave system capable of such rapid injection, sampling, and quenching, that contact times as short as
15 seconds can be achieved. Second, we have developed a rapid but detailed high pressure liquid
chromatographic procedure for classifying coal liquid products into ten fractions of known chemical
functionality.

In this report we will describe the design
and operation of the short contact time reactor and
show its capabilities, and the types of information
we have obtained, by giving the details of several
runs. We will then present some of our early key
findings on the chemistry and kinetics of the solvent-refined coal process. The development and use
of the fractionation procedure will be presented
elsewhere (2).

The type of coal liquefaction method being
studied is exemplified by the solvent-refined coal
process in which a slurry of coal in a solvent derived from the coal in the process is passed through
a reactor in which it is heated in the presence of
H_2 under conditions sufficient to liquefy the coal
and partially desulfurize the products. The unreacted residue is removed and the remainder distilled to produce a variety of products, including the
solvent which is recycled.

II.   COAL CONVERSIONS: TECHNIQUES

A.   Coal Preparation

Coals obtained for study in this project have
been stored in wet lump form in an inert atmosphere.

(2)   Farcasiu, M., Paper in Press, FUEL, Jan. 1977.

All coal preparations were conducted under a sub-
contract with the Department of Aerospace and
Mechanical Sciences, Princeton University, project
titled "Mechanisms of Coal Dissolution and Asphaltene Formation", M. Summerfield, principal investi-
gator. Prior to grinding each sample was placed in
an oven at 125°C for 1 hour in a CO_2 atmosphere and
allowed to cool as the oven cooled. Fragments over
0.5 cm diameter (for which it was assumed no prior
oxidation had occurred) were transferred to a water-
cooled rotary cutter and ground in a CO_2 atmosphere
for from 0.25 to 2 min. All subsequent operations
(including screening and slurry formulation) were
performed in an inert atmosphere.

Portions in various size ranges were obtained
by passing the coal through a series of standard
screens shaken by an electric screen vibrator for 3
hours. Agglomeration was prevented by placing rub-
ber stoppers on each screen. Size distributions
were determined from the yield of coal on each
screen.

The coals used in the runs described in this
paper were West Kentucky 9,14 (a high volatile B
bituminous coal) and Wyodak (a subbituminous coal).
Proximate and ultimate analyses are given in Table
1. Particles ranged in size from ~1 μm to 625 μm
(the majority being 45-150 μm).

B.   Conversion Apparatus and Procedure

A schematic diagram of the high pressure autoclave system is given in Figure 1. The autoclave
was a conventional 300-ml stainless steel autoclave
manufactured by Autoclave Engineering, Erie, Pennsylvania (#MAWP-5400) with fixed head, removable
lower unit, and Magnedrive stirrer; all external
connections were through the head. Connections included 1/4" sample injection line, a 1/8" liquid
sample withdrawal line fitted with a metal filter,
and a 1/16" stainless steel shielded dual thermocouple for reading and controlling temperatures, all
below the liquid level. The stirrer drive was a
Cole Palmer Constant Speed and Torque Control unit
which allowed recording of motor torque during a
run. The stirring blade was spiral shaped and as
large as possible to produce maximum agitation; viscosity was monitored by measuring the stirrer torque.

TABLE 1
Analyses of Coals of the Project
==

Name of Coal	Wyodak	West Kentucky
Mine Location		
State	Wyoming	Kentucky
County	Campbell	Hopkins
Seam	Anderson	9,14
Name of Mine	Bell Ayre	Colonial
Proximate Analysis*		
% Moisture (as rec.)	22.03	6.05
% Ash (as rec.)	3.63	7.83
% Volatile Matter	47.44	36.61
% Fixed Carbon	47.90	55.06
BTU (as rec.)	9599.	12291.
BTU	12311.	13082.
Free Swelling Index	.5	4.
Ultimate Analysis*		
% C	71.82	73.06
% H	5.20	5.00
% O$\neq$	17.12	9.17
% N	.90	1.47
% S (total)	.30	2.97
% S (pyritic)	.06	1.19
% S (organic)	.23	1.36
% S (sulfate)	.01	.42
% Cl	0.0	.00
% Ash	4.66	8.33

*   All analyses are given on a dry weight basis
     unless otherwise stated.

$\neq$ by difference

A cooling coil was mounted inside the vessel in direct contact with the contents. We have incorporated a cooling water reservoir pressured to 200 psi with N_2 to give a high coolant flow rate. The injection system consisted of a barrel with a floating piston insert having "O" ring seals.

To initiate a run a small amount of solvent was pumped up into this barrel from below, then slurry was forced in from below by means of a large metal syringe, followed by a little more solvent so that no slurry remained in the valve. With this sequence, solvent was injected after the slurry to flush all the coal into the vessel. Above the piston was a reservoir of squalane (easily detected by our analytical system in case leaks occurred) which was pressured with N_2 just before the injection. All lines throughout the system were fitted with appropriate vents, rupture discs, drop-out pots, check valves, filters, pressure gauges, etc.

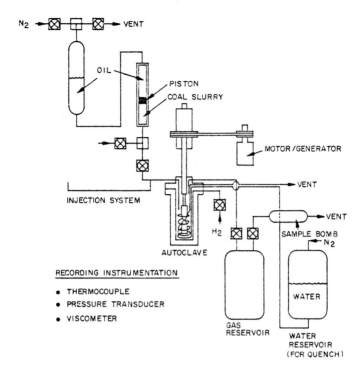

FIG. 1
SCHEMATIC DIAGRAM OF HIGH PRESSURE AUTOCLAVE

In a typical run, 60 g solvent was placed in the vessel, the unit was sealed, flushed twice with H_2, pressure-tested with H_2 at the intended reaction pressure, and vented to 200 psi. Heating with an electric heater and stirring (1200 rpm) were then initiated. As the desired operating temperature was approached (typically after about 1 hr) the injection system was sequentially loaded with 15 g solvent, 40 g 1:1 solvent coal slurry, and 5 g solvent. The time between injector loading and injection was kept as short as possible to minimize slurry settling. If the run was to be very short the H_2 pressure was increased so that the desired pressure was reached after injection, otherwise it was adjusted after injection.

When the temperature lined out 10°C above the desired temperature, a solvent sample was taken from the vessel to establish the pre-injection solvent

composition. The injector was pressured to ~100 psi
above the vessel and the contents forced into the
vessel in about 1 sec. The temperature drop under
these conditions, depending upon the temperature
and the exact amounts of material in the vessel and
injector, was generally 50-115°C. Recovery to the
desired temperature required about .5-1 min; reaction time was assumed to start when the temperature
had recovered to 10°C below the desired reaction
temperature. A typical time-temperature profile is
shown in Figure 2. Liquid and gas samples were oc-
casionally taken during a run; pressure was adjusted if significant drops occurred during sampling.

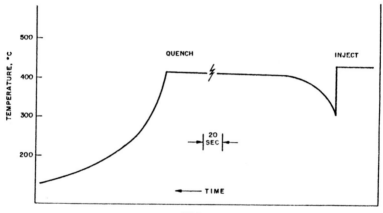

FIG. 2
TYPICAL TIME-TEMPERATURE PROFILE

To end a run, the heater was removed and
simultaneously cooling water flow was started. The
temperature typically dropped about 100°C within
10 sec. The temperature was adjusted to 125°C and
H_2 was added to raise the pressure to ~1200 psi.
The stirring speed was then reduced to ~10 rpm and
the gas vented over a period of 5 min through a
dry ice trap into an 18 l evacuated reservoir. The
trap then contained water, some solvent, and some
light coal products. Water was determined volumet-
rically. The light components were distinguished
from solvent by gas chromatography. The gas reservoir pressure was noted and a sample taken for mass
spectroscopy.
     The autoclave vessel was then cooled to room

temperature and opened. The contents were removed and solids washed out with pyridine; solids adhering to surfaces were loosened with a brush. The mixture of solids and liquid was passed through a Soxhlet thimble; the thimble contents were washed with pyridine which was then combined with the liquid portion. The solids were then extracted with pyridine for 16 hrs; the extract was added to the other liquids. The residue was dried at 125°C in a vacuum oven and stored under Ar until further characterization was carried out.

The pyridine was removed from the combined liquids with a rotary evaporator. The resulting mixture of solvent, SRC, and light organic coal products was subjected to a vacuum distillation at 12 mm pressure. A first cut was taken at 130°C (true boiling point 260°C). If the solvent was our synthetic solvent, this was in the first cut; the boiling point of 2-methylnaphthalene is 240°C. A second cut was taken at 200°C (true boiling point 343°C, 650°F). The residue was the SRC. Within each distillate, product was distinguished from solvent by gas chromatography. The second cut was usually so small that it had to be removed from the condenser with tetrahydrofuron.

The injector system was then flushed with THF to remove uninjected coal which was dried and weighed to complete the mass balance.

Temperature and pressure were recorded throughout the autoclave run on a strip chart.

All slurries for conversions were about 1:1 coal-to-solvent by weight and were made up at least 24 hrs in advance of a run to insure equilibration.

We have recently converted to a new injection system using the same barrel but no piston. Instead, the slurry was driven directly into the autoclave with H_2. Injections were found to be easier and more complete with this method. The slurry was simply poured into the injector, a solvent layer was added, pressure was applied, and the injection made. Washing was as described above.

The products of a run were separated by the work-up into the following fractions: H_2S, H_2O, CO,

CO_2, CH_4, C_2-C_5 (fully-identified by mass spec), C_6-257°F (the contents of the trap, corrected for H_2O and solvent), 257-650°F (the overhead from the distillation, corrected for solvent), SRC (pyridine-soluble, 650°F+), and residue (pyridine-insoluble). Elemental analyses of the SRC and residue were determined by Galbraith Laboratories, Knoxville, Tennessee.

C.   Synthetic Recycle Solvent

In the choice of a reaction solvent, three factors were considered. First, the coal products must be distinguishable from the solvent. Second, the solvent must have, as nearly as possible, properties similar to true recycle solvents in terms of chemical functionality, H-transfer properties, and ability to solubilize products. Third, the solvent must provide a means by which the extent of H-transfer and the occurrence of thermal or catalyzed side reactions (such as cracking and isomerization) could be evaluated.

The following synthetic solvent composition was chosen: ~2% 4-picoline; ~17% p-cresol; ~43% tetralin; ~38% 2-methylnaphthalene. This mixture has the proper amounts of basic nitrogen, phenol, and hydroaromatics. It is more aromatic and lower boiling than a real recycle solvent.

A paper is in preparation describing in detail the reactions of this solvent mixture under coal liquefaction conditions, with and without coal or coal products present. Much has been learned about the chemistry of the SRC process from this system. A brief discussion will be presented here to explain the hydrogen consumption data given in the next section. H-consumptions are given in g/100 g coal.

By following the solvent composition, we can measure the thermal background reactions by observing the isomerization of tetralin to methylindane. This is not affected by the presence of coal. The hydrogen transfer rates are affected by coal, possibly by catalysis by coal minerals. This is measured by observing the formation of methyltetralin from methylnaphthalene (by transfer from tetralin to form naphthalene). Correction can be made for the reaction in the absence of coal, either by calculations based on methylindane formation or by comparison to blank runs with solvent alone. Hydrogen consumption from the solvent by coal or its products is determined by observing the formation of naphthalene and correcting for the hydrogen only transferred to make methyltetralin, and the naphthalene produced through thermal reactions. The solvent: coal ratios are high, and temperatures and H-consumption low. Our reactions are therefore not subject to solvent hydrogen depletion or an approach to thermodynamic equilibria.

III.   COAL CONVERSIONS: PRODUCT DISTRIBUTIONS

Three key factors in coal liquefaction processes for which data and understanding are particularly important are: 1) the rate of conversion to soluble products and the maximum yield obtainable, 2) sulfur removal, and 3) the hydrogen consumption required to achieve a high yield of low sulfur product. [Our results on the chemical nature of the SRC's will be presented elsewhere (2)].

In order to derive the kinetics of the conversion of coal to soluble form, it is necessary to develop understanding of reactions which occur very rapidly during the initial transformations of the coal. Most commercial operations or pilot studies now being operated can provide data only of products that result from extended reactions of coal and some process optimizations may not be observable under these conditions. It has been reported (3) that coal can be dissolved rather rapidly at temperatures ≥800°F. The shortest times that were defined were 2-5 min. It was also noted that the yield of soluble product could actually decline on extended reaction at high temperature due to char formation.

The emphasis of the work described in this report was on developing understanding of the chemical nature of the initially-soluble coal products and mechanisms by which they are formed and interconverted. Our initial efforts have focused on con-

_____

(3)   Monta, M., and Hirosawa, K., Nenryo Kyokai-
      Shi, 53 (564) 263-71, (1974).

version of coal under mild conditions (~800°F), where the rate of formation of char is rather low. This will provide the background information needed to study coal conversion at higher temperatures (charring conditions) which is one of the objectives of future work.

A series of conversions of West Kentucky and Wyodak coals was carried out at 800°F, 1000-1300 psig H_2 in our synthetic solvent. The balances and product compositions of these runs are shown in Table 2. (Disregard computer-generated representation of integers as having two decimal places.)

TABLE 2
Run Balances for Coal Conversions at 800°F in Synthetic Solvent

Run Number	West Kentucky 14				Wyodak	
	10.00	9.00	7.00	12.00	19.00	31.00
Temperature, °F	800.00	800.00	798.00	800.00	806.00	800.00
Pressure, psig H_2	1348.00	1300.00	1160.00	1030.00	1415.00	1072.00
Duration (feed), min	.50	1.30	40.00	417.00	1.30	137.50
MAF Conversion, wt %	50.00	78.20	92.52	96.10	45.97	91.52
Solvent/Coal	4.65	6.66	4.56	9.06	8.48	6.17
H_2S	.24	.20	.67		.01	
Water	4.39	3.46	4.87	5.16	4.21	4.59
CO	.10		.16	.31	.28	1.28
CO_2	.49	.54	.83	1.65	1.61	6.04
C_1	.17	.18	1.18	8.57	.17	2.50
C_2-C_5	2.51	.36	1.60	8.47	.21	4.53
(C_6-257°F)	.34	.14	.22	1.14	.14	.14
(257-650°F)			2.89	9.40	.77	2.38
SRC	46.78	76.12	80.11	61.05	38.53	70.03
MAF Residue	49.76	21.80	7.46	3.93	54.03	8.48
Balance	104.78	102.80	100.00	100.00	99.96	100.00
Ash in Residue	19.30	31.20	53.08	52.50	8.20	37.21
Duration (solv.), min	17.00	15.00	42.00	429.00	32.00	149.50
4-Picoline	1.90	1.90	1.87	3.20	2.01	1.28
P-Cresol	16.10	16.10	16.11	15.80	16.21	13.07
Methylindane	.97	.92	1.88	6.50	1.37	2.59
Tetralin	41.00	41.22	33.41	25.30	40.02	25.22
Naphthalene	2.24	2.10	8.64	14.50	2.32	18.13
Methyltetralin	.09	.28	1.44	4.70	.30	2.37
2-Methylnaphthalene	37.67	37.48	36.62	29.60	37.76	37.36
H_2 Consumption	.20	.24	.89	1.82	.41	2.58
Footnote	5		11	8		10

(5) Gas analysis questionable.
(8) Piston press. resisted syringe; balance forced to total recovered products.
(10) Balance forced; samples removed during run.
(11) Water content based on oxygen balance.

TABLE 3
West Kentucky and Wyodak Coal Conversions

Run No.	Kentucky Coal						Wyodak Coal		
	Ext	10.00	9.00	7.00	12.00	Ext	19.00	31.00
Time (mins.)	.00	.50	1.30	40.00	417.00	.00	1.30	137.50
% Sol.	28.00	50.00	78.20	92.52	96.10	11.50	45.97	91.52
SRC Yield	28.00	46.80	76.10	80.11	61.05	11.50	38.53	70.03
% O SRC	9.47	6.68	7.25	5.15	2.93	12.20	11.75	5.08
% S SRC	2.17	1.51	1.40	1.31	.63	.70	.53	.37
H_2 Consp.	.00	.13	.34	.89	1.59	.00	.41	2.46

It can be seen from the above data that conversion of the coal to >90% soluble form occurs very
rapidly (we estimate ~3 min for West Kentucky and
~20 min for Wyodak).

In order to achieve >90% solubility, the oxygen content of the West Kentucky SRC had only to be
reduced to about 6 wt.%. The oxygen content sys-
tematically became lower with increasing contact
time and there appear to be two forms of oxygen
which are kinetically distinct. One form (~40% of
all the oxygen) is very readily lost. The other
form requires more vigorous treatment to eliminate
it. This is shown in Figure 3, where the log of the
% oxygen remaining in West Kentucky coal products is
plotted against time (this assumes first-order dependence).

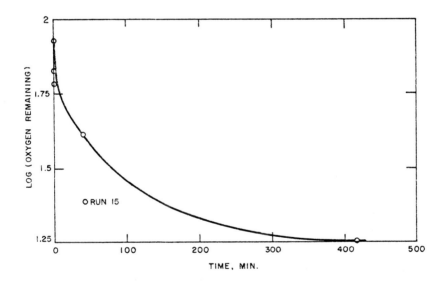

FIG. 3
OXYGEN REMAINING IN WEST KENTUCKY COAL

The <u>organic</u> sulfur content of SRC appears to be
linearly related to the oxygen content of SRC. This
is shown in Figure 4. Again, 40% of the organic
sulfur is easily removed, but the rest is more difficult. Thus, merely dissolving the coal is not
sufficient to lower the sulfur content to acceptable
levels, at least not for West Kentucky 9,14.

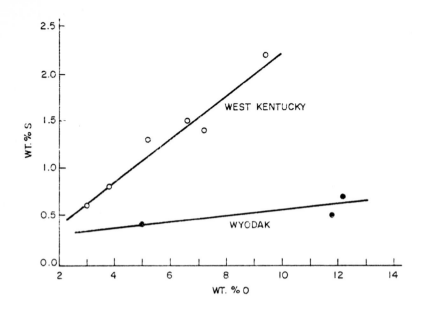

FIG. 4
PERCENT S IN SRC vs. PERCENT O IN SRC

The nitrogen content of all SRC's was not sub-stantially affected by solvent refining of the coal. Thus, catalytic processing may be needed to lower nitrogen content, if this is required for end use.

The time-yield behavior for Wyodak and West Kentucky coals appears to be different. There may, however, be a common feature in that the maximum yield of SRC appears to be co-incident with the minimum time required for >90% conversion to soluble form. This is illustrated in Figures 5 and 6. It can be seen that the yield of SRC from West Kentucky 9,14 coal continually drops after 5-10 min of re-action. This lowering in yield is not due to char for-mation, however, but is the result of converting SRC to solvent and lighter products.

As can be seen in Figure 6, one significant kinetic parameter that can be used as an alternative to time is the percent O converted to CO_2 and H_2O. This parameter is not meaningful when comparing two different coals, however, as the amount and chemical type of oxygen can vary widely.

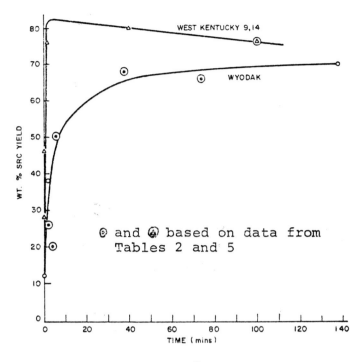

FIG. 5
YIELD OF SRC vs. TIME

The hydrogen consumption observed for a given coal generally increases with increasing degree of solubilization. However, the yield of SRC vs. hydrogen consumption goes through a maximum. This is shown in Figure 7. Solvent yields become appreciable only with high hydrogen consumption and methane formation appears to be a major factor in high hydrogen consumption at high conversion levels.

Only limited data are presently available on the effect of temperature on the reaction kinetics or selectivity. This aspect of the work is presently being pursued. One comparison was made of the conversion of West Kentucky 9,14 coal at 800°F and 850°F at 30 sec contact time. The product balances are shown in Table 4.

It can be seen that even at this short contact time, major changes in the product are observed on increasing the temperature by ~50°F. The overall conversion increased by 20% and the yield of SRC

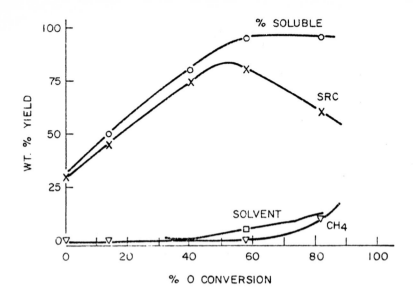

FIG. 6
PRODUCT YIELDS vs. PERCENT OXYGEN CONVERSION
FOR WEST KENTUCKY COAL

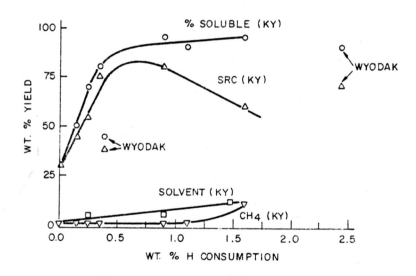

FIG. 7
YIELD OF SOLUBLE PRODUCTS, SRC, SOLVENT, AND
CH4 vs. H CONSUMPTION 800°F ~ 1500 PSI. H2

TABLE 4
The Effect of Temperature on Liquefaction of West Kentucky Coal

Run Number	10.00	14.00
Solvent	Synth	Synth
Coal	WKy 14	WKy 14

Temperature, °F	800.00	847.00
Pressure, psig, H_2	1348.00	1012.00
Duration (feed), min.	.50	.50
MAF Conversion, wt. %	50.00	70.07
H_2S	.24	.10
Water	4.39	3.65
CO	.10	.21
CO_2	.49	.83
C_1	.17	1.25
C_2-C_5	2.51	2.71
$(C_6-257°F)$	.34	.63
(257-650°F)	.00	4.38
SRC	46.78	56.31
Residue	49.76	28.93
Balance	104.78	100.00
Ash in Residue	19.30	16.93
Duration (Solv), Min.	17.00	15.00
4-Picoline	1.90	2.14
p-Cresol	16.10	16.98
Methyl Indane	.97	1.93
Tetralin	41.00	39.55
Naphthalene	2.24	1.82
Methyl Tetralin	.09	.26
2-Methyl Naphthalene	37.67	37.31
FOOTNOTE:	5.00	7.00

(5) Gas Analysis questionable.
(7) Mat bal. low; bal forced to total recovered prods; gas
analysis poor.

was increased by about 10%. Increased solvent range
material yields were also observed. Because the
oxygen conversion was still below 40%, no additional
hydrogen consumption was noted.

A brief series of runs was made to examine
the relative importance of H_2 and of solvent H-donor
capacity in the early stages of liquefaction. The
results are presented in Table 5. It is apparent
that in the absence of H_2 gas, and H-donor solvent,
heating the coal in pyridine to 800°F under N_2 pressure for either 1.3 or 60 min did not increase the

TABLE 5
Criticality of Solvent and Hydrogen

Run Number	16.00	9.00	17.00	18.00	7.00	3.00
Solvent	SYNTH	SYNTH	PYR	PYR	SYNTH	PYR
Coal	WKY 14	WKY 14	WKY 14	WKY 14	WKY 14	WKY 14
Temperature, °F	800.00	800.00	800.00	800.00	798.00	796.00
Pressure, psig H_2	.00	1300.00	.00	809.00	1160.00	.00
Duration (feed), min	1.28	1.30	1.28	1.18	40.00	60.00
MAF Conversion, wt %	65.36	78.20	30.95	40.13	92.52	29.78
Solvent/Coal	5.04	6.66	5.17	5.10	4.56	5.00
H_2S	1.00	.20			.67	
Water	5.28	3.46	4.75	4.59	4.87	
CO				.05	.16	
CO_2	1.90	.54		.05	.83	1.40
C_1	.53	.18			1.18	1.32
C_2-C_5	2.43	.36		.05	1.60	2.01
(C_6-257°F)	.16	.14	.11		.22	
(257-650°F)	.58		.05	.48	2.89	12.11
SRC	58.58	76.12	26.00	28.56	80.11	19.67
MAF Residue	37.35	21.80	69.05	59.87	7.46	63.49
Balance	107.81	102.80	100.00	93.65	100.00	100.00
Ash in Residue	20.63	31.20	10.97	12.30	53.08	12.02
Duration (solv.), min	21.00	15.00	18.00	22.00	42.00	60.00
4-Picoline	1.78	1.90			1.87	
P-Cresol	15.43	16.10	.13	.02	16.11	
Methylindane	1.79	.92			1.88	
Tetralin	38.95	41.22	.37	.25	33.41	
Naphthalene	2.86	2.10	.02		8.64	
Methyltetralin	.12	.28			1.44	
2-Methylnaphthalene	39.08	37.48	.27	.18	36.62	
H_2 Consumption	.29	.34			.89	
Footnote					11	2

(2) Balance forced; feed not injected.

(11) Water content based on oxygen balance.

pyridine solubility (Soxhlet extraction) above that observed for untreated coal. From Run 18 it can be seen that the use of H_2 without a H-transfer agent resulted in a small increase in solubility. In Run 16 the use of a H-donor solvent without H_2 gave a substantial increase, approximately doubling the conversion over extraction alone. For comparison, use of H_2 and H-donor gave 2.6 times the conversion.

These results suggest that in the early stages of liquefaction the availability of a H-transfer agent is the most important factor, although H_2 gas does have an effect. The possible importance of mass transfer of liquid solvent and product molecules to and from the reactive surface is clearly indicated. Complete liquefaction of Kentucky 9,14 coal in the absence of H_2 appears feasible in the presence of a good H-donor.

IV.   INITIAL PRODUCTS IN SOLUTION

Information has been obtained on two additional facets of coal liquefaction: 1) the nature of the first solubilized species, and 2) the possibility of mass transport limitations on escape of the initial products from the particles.

To investigate these areas, we needed a procedure for distinguishing between coal that has been dissolved and is reacting but still remains within particles, and species that have escaped to the bulk solvent. Accordingly, we have attached a filter, below the liquid level inside the autoclave, to the liquid samplingline. We can, therefore, withdraw solution samples during a run, and these samples can contain only that portion of SRC product that is outside the coal particles. (This is also useful in scoping conversion vs. time to establish appropriate quench times for other runs.) The SRC content of these samples was determined by high pressure liquid chromatography. We used purified SRC's from previous high and low conversion runs to calibrate the system. A highly-polar solvent, pyridine, and a non-activated, trimethylsilylated silica, were used to ensure complete elution of SRC. Corrections were made for the synthetic solvent, which gives a very low response to our moving hot wire detector.

The results of run AC-31 (Wyodak coal, 800°F) are reported in Table 6. (The observations of molecular weight distribution changes as determined on these samples by gel permeation chromatography will be discussed below.)

We find that at least in the first four minutes of Run AC-31 there is less SRC observed in solution than we know had been produced. In Run AC-19, 1.3 min long, 38.5% of the coal was converted to SRC; after ~3 min in AC-31 only 2/3rds of this amount was observed in solution. It appears, therefore, that in the initial stages of the reaction a significant portion of the product remains within the coal particles. It must be pointed out that the SRC in Run AC-19 was obtained after exhaustive pyridine extraction. The above conclusion is valid unless we make the unexpected finding that the synthetic solvent at 800°C is a poorer solvent for SRC

TABLE 6
Wyodak SRC in Solution During Run AC-31

| Time, Min | Weight Percent of Original Coal | | Total |
	Low Molecular Weight (300-900 MW)	High Molecular Weight (>2000 MW)	
1.20	16.86	10.97	27.80
3.60	15.41	6.13	21.53
6.00	44.32	8.48	52.80
19.50	44.99	6.70	51.69
38.00	65.77	6.87	72.64
74.00	66.45	4.79	71.24
137.50	70.54	4.03	74.57

than pyridine at typical extraction conditions (<100°C in our apparatus).

There is, therefore, a strong indication that even in a fast-stirred reactor with a good H-transfer solvent, there may be reactions occurring inside coal particles that could be different from those occurring outside. The results of initial studies indicate that SRC contains a bimodal molecular weight distribution. There is a high molecular weight component (>2000 MW), or very strongly associated complexes of smaller very functional molecules, which in the early stages represent up to 40% of the product appearing in solution. The relative concentration of this material then rapidly declines, producing material in the 300-900 molecular weight range. This is shown in Figure 8. For the conversion of Wyodak coal in which the contents of the autoclave were sampled with time and analyzed by GPC, the results are shown in Table 6. These show that the absolute yield of this high molecular weight material (weight % of the coal fed) decreases throughout the run. Liquid chromatographic study of the high molecular weight material shows that it is very polar and highly functional. Furthermore, we find that the THF soluble portion of the pyridine extract of West Kentucky coal shows none of this high molecular weight material; the THF insoluble portion of the extract contains a significant amount of high molecular weight material. In addition, GPC examination of SRC's from our autoclave runs with Kentucky coal show high molecular weight mate-

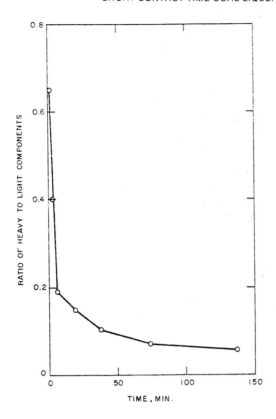

FIG. **8**
WYODAK SRC COMPONENTS

rial in low conversion runs, but very little in
high conversion runs. The key point is that on pyridine extraction or at very low conversion, the
coals show significant amounts of high molecular
weight materials which are consumed rapidly on further conversion (times greater than ~3 min).

V.   CONCLUSIONS AND WORK IN PROGRESS

From the data presented here we conclude that
coal dissolution is very fast and requires very
little hydrogen consumption. The presence of H_2
gas in the early stages of conversion is not critical, but a good H-donor solvent must be present.
Sulfur and oxygen are removed in a kinetically-
parallel fashion. About 40% of each may be removed

readily and rapidly with little or no H-consumption; thereafter, considerably more hydrogen is consumed than the stoichiometry requires for the production of H_2S and H_2O. The initial products of coal dissolution contain significant amounts of high molecular weight material which is rapidly converted to low molecular weight products. The highest SRC yield is obtained early in the reaction process; improvement of SRC quality is accompanied by a decrease in yield and a large increase in hydrogen consumption.

Our present work includes more detailed chemical analysis of the SRC, extension of our studies to more coals and a wider range of conditions, and an investigation of possible mass transport limitations.

ACKNOWLEDGEMENTS

We wish to acknowledge helpful discussions with M. Farcasiu and the experimental assistance of B. O. Heady.

REFORMATION OF INORGANIC PARTICULATES SUSPENDED
IN COAL DERIVED LIQUIDS AND IMPROVED SEPARATION

S. Katz and B. R. Rodgers
Oak Ridge National Laboratory

The inorganic constituents of coal are found as very small suspended particulates after conversion of the coal to liquid product by pyrolysis or hydrogenation. Separation of the particulates by filtration, hydrocloning, and centrifugation has at best been marginally satisfactory. Reported here are techniques to change the particulates' size and shapes, and thereby improve the subsequent separation. Treatments involving heat, solvent fractionation, and additives are described and limiting conditions for their applications are discussed.

For coal liquefaction processes, the separation of fine inorganic particulates from the organic product matrix represents an important technical problem. This problem is difficult because the matrix is viscous and many of the particles are in the micron and submicron size range. Of the many separation methods and devices investigated, the following have shown some success in separating the particles: hydroclones, centrifugation, magnetic separation, solvent extraction, solvent fractionation, and filtration. All of these, however, suffer in varying degrees from unsatisfactory operation, maintenance, throughput, and cost. To help correct these shortcomings, we have undertaken a study of techniques by which the fine particulates might be enlarged or more effectively collected. Some progress toward that end has been described in the open and patent literature for solvent addition. In these works, two mechanisms by which solvent improves separation have been suggested: (1) the solvent may soften or dissolve a coating from the particles, permitting physical attractive forces to cause agglomeration; or (2) the solvent may cause phase separation--the heavier phase acting as a collection flocculant.

I.   THE LABORATORY SEDIMENTATION TESTS

The laboratory sedimentation tests were conducted in a vertical 18-inch-tall by 1-inch-diameter metal tube using unfiltered oil from the Solvent Refined Coal process. Normally, 175 grams of this oil containing suspended particles was placed in the tube. The sealed tube was quickly heated to test temperature and then cooled rapidly at the end of the test. The contents, as shown in Fig. 1, were then separated into 10 or 11 fractions for examination by filtration, microscopy, or chemical analysis. Filtration was achieved by the use of a laboratory technique described in the July 1976 issue of Industrial and Engineering Chemistry, Process Design and Development.

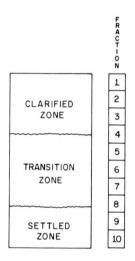

Fig. 1.   Settling zone identification.   Generally:
(1) Clarified zone contains particles less than 5 μ and has a total solids concentration much less than the UFO.   (2) Transition zone contains a few particles up to 15 μ and varies from a low concentration to that of the UFO.   (3) Settled zone contains nearly all large dark particles and has a concentration much greater than UFO.

II.   EFFECTS OF SOLVENT DILUTION ON SETTLING

The solvent dilution tests, using toluene or SRC recycle solvent as the diluent, gave comparable results; an appreciable improvement in settling was obtained at 20% dilution.

Table 1 shows data for settling at 310°C for 1 hour with 5, 10, and 20% dilution with recycle solvent; the ash for the 20% dilution is approaching the EPA requirement of 0.15% in the top 70% of the sedimentation tube. The role of particle agglomeration appears well-substantiated here since the agglomerates formed can be observed by optical microscopy to contain as many as 100 of the particles. Calculations of effective particle size from the settling rates indicate that the settling particles are a minimum of 53 microns in diameter, whereas 90% of the particles in the starting test are less than 1 micron in diameter.

TABLE I

Analytical Results for Settling of Process Recycle Diluted, Unfiltered Oil from the SRC Process.

Condition	Fraction No.	Ash (Wt %)	Sulfur (Wt %)
1 hr at 310°C, 5% dilution	1	1.67	0.63
	4	1.52	0.70
	7	1.44	0.67
	11	3.32	0.70
1 hr at 310°C, 10% dilution	1	1.12	0.59
	4	0.72	0.60
	7	1.22	0.67
	10	3.49	0.75
1 hr at 310°C, 20% dilution	1	0.33	0.50
	4	0.24	0.54
	7	0.03	0.54

III. SETTLING WITH HEATING BUT WITHOUT SOLVENT DILUTION

Keeping in mind that with 20% solvent dilution, adequate clarity can be approached by settling for 1 hour at 310°C, let us consider what can be done with heat alone. Figure 2 presents the results of a series of sedimentation tests conducted at temperatures of 100 to 350°C for 1 to 21 hours. The filtration improvement factor is a term describing the relative filter-ability of the third fraction from the top; a factor of 1 is no improvement, whereas a factor of 20 or more indicates clarity meeting EPA standards. Little settling was found after 18 hours at 100 and 150°C. At 200 and 250°C, settling was more rapid and

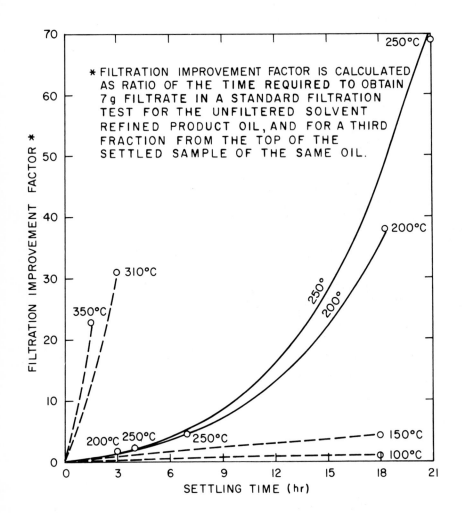

Fig. 2. Improvement in filterability as a function of settling time and temperature.

clarity was reached before 18 hours. The results were good at 310°C in 3 hours and at 350°C in 1 hour. At higher temperatures, settling was poor because of decomposition. Therefore, 300 to 350°C appears to be the best working range. Settling at 350°C seems at least as satisfactory as with 20% solvent dilution at 310°C. The effects on the particles may be considerably different since both the size and distribution of particles apparently change following thermal treatment. Figure 3 shows particles in the starting oil at a magnification of 24,000X. Some of the particles are only 0.01 micron in diameter; the larger particles have rough edges, which suggests that they may be composed of smaller particles.

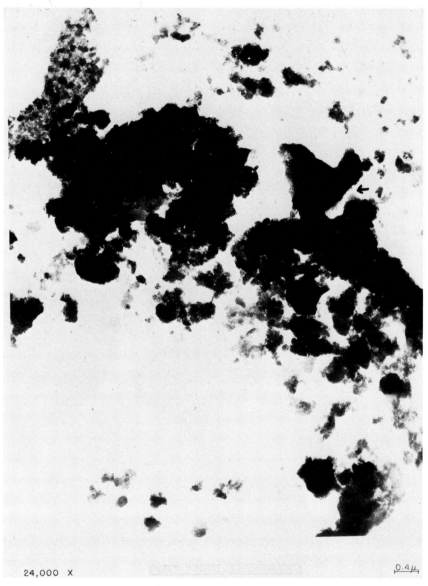

24,000 X   0.4μ

Fig. 3.   Unfiltered SRC oil.

Figure 4 reveals a similar magnification of the particles after treatment at 300°C for 1 hour; few particles less than 1 micron in size remain, and the surface edges appear to be smooth.

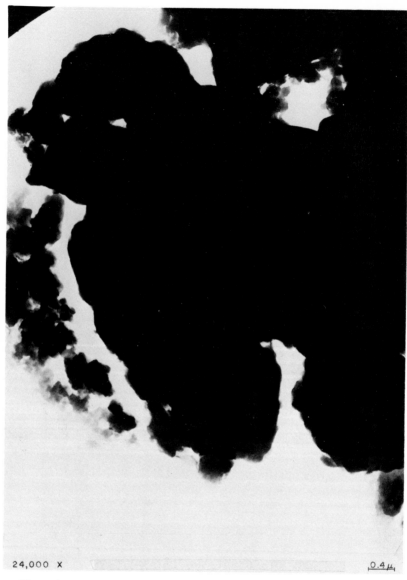

24,000 X   0.4 μ

Fig. 4. Heat treated unfiltered SRC oil.

I. THE LABORATORY SEDIMENTATION TESTS

   The laboratory sedimentation tests were con

appeared to be most effective at higher concentrations such as 10,000 ppm but also seemed to be effective at temperatures as low as 200°C--phase separation in the matrix material and possibly flocculation were noted even at this low temperature.

These laboratory data indicate that, through heat treatment in the presence of a small amount of a promoter, settling takes place in less than 1 hour and, therefore, does meet clarity standards. Other arrangements, such as to couple settling with a polishing step, may be useful. The effect of heat on settling has been confirmed in a larger-scale system; the effect of additives is still being studied. These results will be reported at a later date.

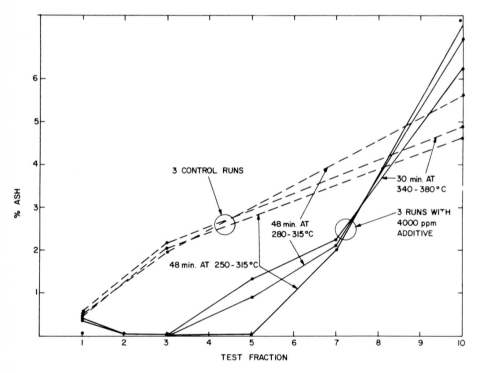

Fig. 5. Comparative settling tests.

V. MECHANISMS OF SETTLING

Table II lists some possible mechanisms which may promote settling. Since many particle types and compositions exist in a suspension in a variable matrix, several settling mechanisms may be operating simultaneously. Agglomeration is known to

TABLE II

Possible Events Leading to More Rapid Settling
===

Initial Events	Final Events
Neutralization of zeta potential	Agglomeration
Melting away of hard organic surface layer	Particle enlargement through recrystallization
Chemical conversion of surface	Flocculation
Phase separation	

occur both after solvent dilution and upon the application of heat. The extent to which charge, melting, or chemical conversion occurs to promote agglomeration has not yet been established. Electron-microscope studies of the material made before and after heat treatment indicate that particle reformation does occur. It implies some solubility of the inorganic species in the matrix organic materials at the test conditions, which is not particularly surprising in view of the polar components of the organic matrix. None of the promoters, when used alone, appears to flocculate and gather up particles as we would expect to observe in aqueous systems; however, some of the additive tests, particularly those with phosphoric acid, phosphoric anhydride, and sulfuric acid seem to cause a small amount of a matrix phase to separate out and collect particulates. The normal changes in viscosity that are induced by temperature or by dilution with solvent are significant but do not account for the measured settling rates. Phase separations, as a result of treatment, can affect viscosity and, subsequently, the settling rates.

COAGULATION AND FILTRATION OF SOLIDS FROM LIQUEFIED COAL OF SYNTHOIL PROCESS

John O. H. Newman, Sayeed Akhtar[a], and Paul M. Yavorsky
Energy Research and Development Administration,
Pittsburgh Energy Research Center

ABSTRACT

The solids in the liquefied coal from the SYNTHOIL process were agglomerated by blending the liquefied coal with a process-derived light oil and holding the blend at 105° - 115° C for 10 minutes in a vessel fitted with a reflux condenser. The agglomerated solids were filtered and the filter cake was washed with a quantity of the same light oil as used for blending. The light oil added during blending and cake-washing was then distilled off from the filtrate to recover a product oil low in solids. Thus, a sample of liquefied coal containing 4.6 per cent ash and 8.1 per cent organic benzene insolubles was converted to product oil containing 0.01 to 0.05 per cent ash and 0.2 to 2.5 per cent organic benzene insolubles by the method. The residual oil in the filter cake was negligible. In a continuous operation, the light oil distilled off from the filtrate will be recycled for blending with liquefied coal and cake-washing.

I. INTRODUCTION

The idea of improving the separation of solids from coal liquefaction product by dilution with a solvent or coagulation with an antisolvent is not new. The literature on the subject is reviewed in a recent report (1). A patent dated 1970 suggested using a solvent such as acetone, or aromatic or aliphatic hydrocarbons (2). An earlier patent suggested adding a paraffinic liquid (antisolvent) to precipitate benzene insolubles and asphaltenes (3). A 1972 patent described the agglomeration of solids by recycling a fraction of the coal

---

[a] To whom all communications concerning this publication are to be addressed.

liquefaction product (4). A 1974 patent, assigned to The
Lummus Company, emphasized the use of an antisolvent with
paraffinic characteristics (5). Additional information on the
agglomeration and separation of solids by the Lummus process
was presented in 1975 (6). Separation of the agglomerated
solids is conducted by gravity settling. Deasphalting of coal
extract with saturated hydrocarbons is described in (7).

The purpose of the present investigation was to coagulate
the solids in liquefied coal from the SYNTHOIL process with a
process-generated antisolvent and to remove the agglomerated
solids by filtration rather than settling. In the SYNTHOIL
process for converting coal to a nonpolluting utility fuel oil,
coal is liquefied and hydrodesulfurized catalytically by
reaction with hydrogen in a turbulent-flow, packed-bed reactor
(8). The gross liquid product is then centrifuged, or filtered,
to remove the minor amounts of residual unreacted coal and
mineral matter. If a method for agglomerating the finely
divided solids in the gross liquid product could be developed,
the subsequent separation of solids will be easier and more
complete. Such a method and the experimental data in support
of it are described below. It is not yet possible to comment
on the economics of the method.

II. METHOD

The method for agglomerating and separating the solids
present in the gross liquid product consists of the following
steps:

1. Prepare a blend of the gross liquid product with an
    accumulation of light oil derived from the SYNTHOIL
    process (9). The blend may contain 40 to 50 weight
    per cent of the gross liquid product.
2. Hold the blend at 105° - 115° C for 10 minutes in a
    vessel fitted with a reflux condenser.
3. Filter.
4. Wash the filter cake with the same light oil as used
    for preparing the blend.
5. Recover light oil from the filtrate by distillation.

In the experiments described below, product oils containing less than 0.1 per cent ash were obtained by the method
and the residual oil in the solids appeared to be negligible.

III. MATERIALS AND EXPERIMENTAL PROCEDURE

Analyses of the gross liquid product and the light oil used in this work are given in tables 1 and 2, respectively. Although not evident from the analysis, the light oil had an ammonia odor. Using a Waring blender, mixtures of the gross liquid product and the light oil were prepared by the following techniques:

1. The components were blended for 5 minutes during which time the frictional heat raised the temperature of the mixture to 60° C.
2. The components were blended for 15 minutes during which time the mixture boiled due to the frictional heat.
3. A mixture of 80 parts by weight of the gross liquid product and 20 parts by weight of the light oil was blended for 5 minutes and then diluted to mixtures containing 10 to 50 per cent by weight of gross liquid product by shaking with calculated quantities of the light oil.
4. The diluted mixtures prepared in 3 were refluxed for 10 minutes, at 105° -115° C.

The mixtures prepared in 1 to 4 were cooled to room temperature for filtration tests and viscosity and density determinations.

TABLE 1

	Weight per cent
Organic benzene insolubles	8.1
Asphaltene[1]	26.3
Oil[2]	61.0
Ash	4.6

[1] *Asphaltene is soluble in benzene but insoluble in pentane.*

[2] *Oil is soluble both in benzene and pentane.*

TABLE 2

Analysis of Light Oil

ASTM distillation D-158

Volume per cent distilled	Temperature, °C
F. D. ------------	78.0
5 ------------	104.5
10 ------------	113.5
20 ------------	124.0
30 ------------	134.5
40 ------------	146.5
50 ------------	158.0
60 ------------	169.0
70 ------------	179.5
80 ------------	193.0
90 ------------	213.0
95 ------------	229.0
End point ----	232.0

Recovery: 96 per cent
Residue: 3 per cent
Loss: 1 per cent

Mass spectrometric analysis

Component	Weight per cent
Saturates, $C_5 - C_{16}$ --------------	44.0
Benzenes -------------------------	21.3
Indanols and benzothiophenes -----	0.2
Phenols -------------------------	7.9
Dihydrophenols, resoricinols, and thiophenols -----------	7.1
Acenaphthenes and biphenyls ------	0.5
Naphthalenes ---------------------	4.5
Indenes -------------------------	1.6
Indans and tetralins ------------	12.7
Total	99.8

The filtration equipment, shown in figure 1, consisted of an 8 inch length of 1-3/8 inch ID steel pipe fitted with a screwed top gas and thermocouple adapter and a bottom filter assembly. The latter comprised a paper disc clamped between a 1.3 inch ID Teflon ring and a disc of 62 x 62 per inch square weave gauze on a perforated brass disc. The filter assembly was tightened with a wrench but the top assembly was only hand-tightened, enabling it to be rapidly opened, refilled, and screwed tight. The filter was clamped above a beaker mounted on a triple beam scale. For filtration rate determinations, mixtures were poured into the tube, the top assembly fitted, a constant nitrogen pressure of 25 psig applied, and the time intervals for collecting known quantities of the filtration noted. With dilute mixtures, it was necessary to build-up filter cake to reduce the filtration rate before rate measurements could be made. Specific resistances of filter cakes were calculated by the method described by McCabe and Smith (10).

The viscosities of the mixtures and filtrates were measured on 8 ml samples with a Brookfield Synchro-Lectric model LVT viscometer in a Thermosel constant temperature unit. The spindle model was SC4-18. The measurements were conducted at room temperature and at a high shear rate of 79.2 sec^{-1} at 60 rpm. A series of viscosity determinations on the gross liquid product over a range of shear rates at three different temperatures had shown that the apparent viscosity is very sensitive to shear rate at low values of the latter (figure 2).

IV. RESULTS

The filtration rates are given as plots of reciprocal filtration rate, $\Delta t/\Delta v$, against average filtrate volume, $\bar{v}$, in figures 3, 4, and 5. Specific cake resistances calculated from the slopes of the graphs in figure 5 are given in table 3. The filtrates obtained from the experiments shown in figure 5 were analyzed for ash, organic benzene insolubles, asphaltene, and pentane-soluble oil and the concentrations of these components in light oil-free SYNTHOIL product were calculated by allowing for the know quantity of light oil in the filtrate. The calculated analysis of the recovered SYNTHOIL product is given in table 4. The viscosities of mixtures of gross liquid product and light oil refluxed for 10 minutes, and of the filtrates obtained from them are given in table 5 and plotted logarithmically against log light oil fraction in figure 6.

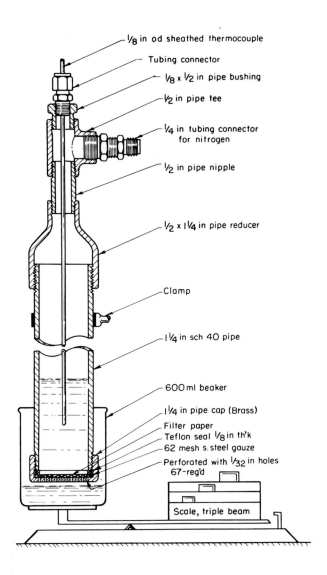

Fig. 1. - Apparatus to measure filtration rates.

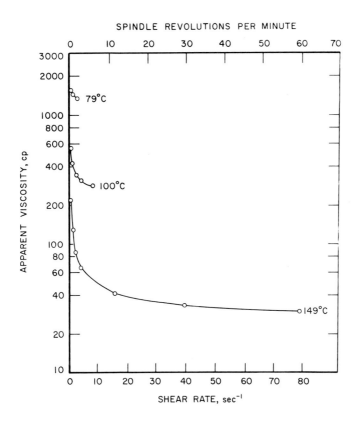

Fig. 2. - *The effect of shear rate on apparent viscosity of gross liquid product.*

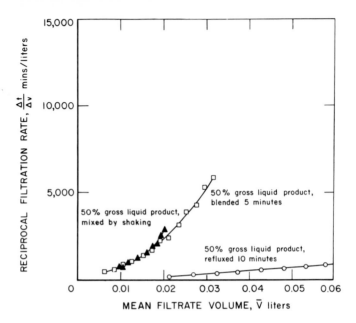

Fig. 3. - *Dependence of reciprocal filtration rate on mixture treatment, (50% gross liquid product, 50% light oil).*

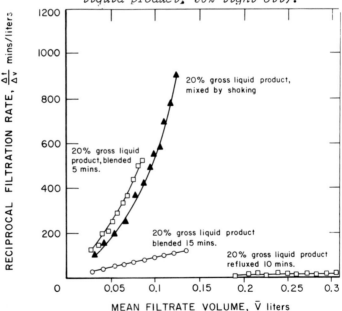

Fig. 4. - *Dependence of reciprocal filtration rate on mixture treatment, (20% gross liquid product, 80% light oil).*

TABLE 3

Specific Resistances of Filter Cake from Mixtures of Gross Liquid Products and Light Oil Refluxed for 10 Minutes

Gross liquid product in mixture, weight per cent	Specific cake resistance	
	$ft/lb \times 10^{11}$	Technical units
10	$\underline{1}/$	$\underline{1}/$
20	0.1	0.06
30	0.1	0.07
40	0.5	0.3
50	3.2	2.0
60	$\underline{2}/$	$\underline{2}/$

[1] Filtration rate was too rapid for rate measurement.

[2] Filtration rate was too slow for rate measurement.

TABLE 4

Analysis of Light Oil-Free SYNTHOIL Product Recovered
From Filtrate (a)

Gross liquid product in mixture, weight per cent (b)	Analysis of Recovered SYNTHOIL Product			
	Ash	Organic Benzene Insolubles	Asphaltene	Oil
10	0.03	0.2	27.0	72.8
20	<0.02	0.4	24.3	75.3
30	0.04	0.4	26.8	72.8
40	0.05	1.4	29.5	69.1
50	0.01	2.5	31.3	66.2

(a) Calculated from analyses of the filtrate and the known quantities of light oil in them.

(b) The mixtures were refluxed for 10 minutes.

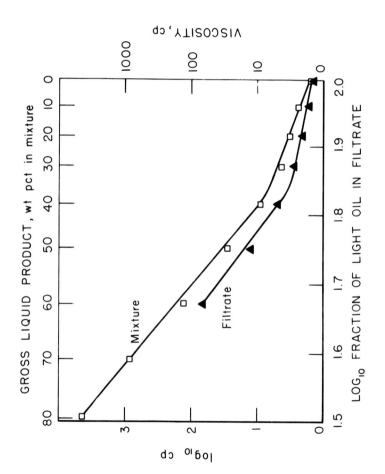

Fig. 6. – The effect of light oil concentration on viscosity of mixture and filtrate

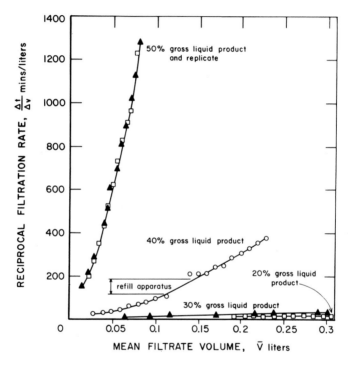

Fig. 5. - *Dependence of reciprocal filtration rate on gross liquid product concentration in light oil (refluxed).*

TABLE 5

Viscosity of Mixtures and Filtrates[a]

Gross liquid product in mixtures, weight per cent	Visocosity of mixture, cp	Viscosity of filtrate, cp
0 (neat light oil)	–	1.6
10	2.5	1.8
20	3.4	2.2
30	4.6	3.0
40	8.6	5.1
50	27	13
60	129	65
70	800	N. D.[b]
80	4,090	N. D.[b]

[a] Mixtures refluxed for 10 minutes

[b] N. D. = Not determined

V. DISCUSSION

Generally, the results suggest that the gross liquid product contains fine solids that block the capillaries of filter cakes and prevent rapid filtration. The solids are agglomerated by thermal treatment with light oil and cooling to room temperature. High concentrations of light oil give very open filter cakes capable of extremely rapid filtration rates. Addition of light oil also reduces the viscosities of the gross liquid products and filtrates.

Figure 3 shos that a 50 per cent mixture of gross liquid product in light oil prepared without sufficient heating was slow to filter. A marked improvement in filtration rate was obtained by refluxing the slurry for 10 minutes, where a specific cake resistance of 3 x 10^{11}ft/lb (2 technical units) was

found. Figure 4 shows similar effects with a 20 per cent mixture of gross liquid product in light oil. Blending for 5
minutes gave a cake resistance little different from that of
the shaken mixture. Blending for 15 minutes showed a marked
improvement but not as great as that obtained by refluxing the
mixture for 10 minutes, where a apecific cake resistance of
0.1×10^{11} ft/lb (0.06 technical units) was attained.

Comparison of the reciprocal filtration rates of 50 per
cent to 20 per cent mixtures of gross liquid product in light
oil, as seen in figure 5, highlights the marked dependence of
the filtration rate of refluxed mixtures on concentration.
The filtration rate of the 10 per cent mixture was too rapid to
measure and that of the 60 per cent mixture was very slow.
The specific cake resistances found in these tests and given
in table 3 tend to a lower limit of 0.1×10^{11} ft/lb (0.06
technical units) at 30 per cent gross liquid product and less.
The curvature of the plots for the 50 per cent and 40 per cent
gross liquid product concentrations shows that, inspite of
agglomeration by thermal treatment, capillary blockage did
occur. Once capillaries blocked, they could not be opened by
washing.

The results in table 4 show that the concentration of
organic benzene insolubles in the recovered SYNTHOIL product
is also markedly reduced by thermal treatment with light oil
followed by filtration. The odd breakpoint in the plots of log
viscosity verus log concentration of gross liquid product in
mixtures shown in figure 6 could be due to precipitation of
the organic benzene insolubles and larger asphaltene molecules
(macromolecules) normally soluble in the products of coal
liquefaction. The breakpoint occurs at about 30 per cent to
40 per cent gross liquid product concentration where there is
a sharp decrease in organic benzene insolubles in the recovered SYNTHOIL product.

It should be noted that no solid filter-aids were added
in these tests, the agglomerated solids acting as their own
filter medium. From the electron micrographs in figures 7 and
8, the agglomerates appear to be chain-like. Many fine particles are also clearly visible.

The agglomerating action, or the antisolvent characteristics, of the light oil may be due to the saturates in it.
As given in table 2, mass spectrometric analysis of the light
oil showed it contained 44 per cent of C_5 to C_{16} saturates.

Fig. 7. - Electron micrograph of filtered solid.

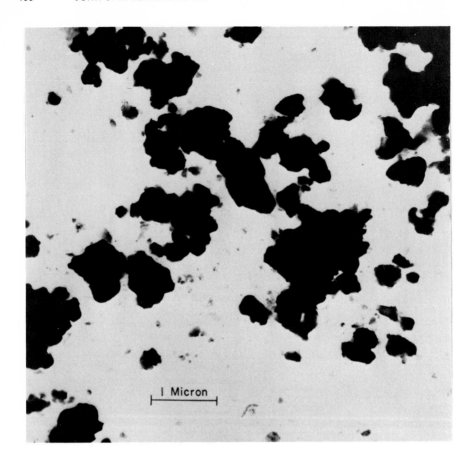

Fig. 8. - Electron Micrograph of Filtered Solid.

VI.   CONCLUSIONS

1. The finely divided solids in the gross liquid product from the SYNTHOIL process are coagulated by heating a mixture of the gross liquid product and a process-derived light oil at 105° - 115° C for 10 minutes.

2. The coagulated solids are amenable to rapid separation by filtration after the mixture is cooled.

3. The concentration of ash in the product oil is not sensitive to the concentration of the gross liquid product in the mixture (see table 4).

4. The filtration rate of a mixture is sensitive to the concentration of the gross liquid product in it (see table 3). The filtration rate of a mixture containing 60 per cent gross liquid product was too slow for practical purposes. The mixture must contain a minimum of 50 per cent light oil for adequate rate of filtration.

VII. ACKNOWLEDGEMENTS

This work was in part supported by the British Coal Utilization Research Association Ltd. through the award of the Jack Carrington Memorial Fellowship to Mr. John O. H. Newman. The work was conducted at the facilities of the Pittsburgh Energy Research Center. The encouragement of the Center's Director, Dr. Irving Wender, is gratefully acknowledged.

VIII. REFERENCES

1. M. S. Edwards, B. R. Rodgers, and R. Salmon. Coal Technology Program Supporting Research and Development on Separations Technology: Phase I Report, Oak Ridge National Laboratory, TM - 4801, March 1975.

2. E. S. Johanson, C. S. Schuman, H. H. Stotler, and R. H. Wolk. U. S. Patent 3,519,533, July 1970.

3. E. Gorin. U. S. Patent 3,018,241, January 23, 1962.

4. R. J. Fiocco and E. L. Wilson. U. S. Patent 3,687,837, August 29, 1972.

5. M. C. Sze and G. J. Snell. U. S. Patent 3,856,675, December 24, 1974.

6. M. C. Sze and G. J. Snell. A New Process for Removing Ash from Coal Liquefied by Hydrogenation. Presented at American Power Conference, Chicago, Illinois. April 21-23, 1975.

7. E. Gorin, C. J. Kulik, and H. E. Lebowitz. ACS Division of Fuel Chemistry Preprints, Vol. 20, no. 1, 1975. p. 79

8. Sayeed Akhtar, Nestor J. Mazzocco, Murray Weintraub, and Paul M. Yavorsky. SYNTHOIL Process for Converting Coal to Nonpolluting Fuel Oil. Energy Communications, vol. 1, no. 1, 1975. pp. 21 - 36.

9. Sayeed Akhtar, James J. Lacy, Murray Weintraub, Alan A.
    Reznik, and Paul M. Yavorsky. The SYNTHOIL Process –
    Material Balance and Thermal Efficiency. Presented at the
    67th Annual AIChE Meeting held at Washington, D. C.,
    December 1 – 5, 1974.

10. W. L. McCabe and J. C. Smith. " Unit Operations of Chemical
    Engineering", McGraw Hill Book Company, 1967.

FILTERABILITY OF A COAL-DERIVED LIQUID

Murray Weintraub, Milton J. Weiss,
Sayeed Akhtar and Paul M. Yavorsky
U. S. Energy Research and Development Administration

ABSTRACT

Solids separation is an important phase in all coal liquefaction processes. It is a difficult step because of the small particle size, wide size distribution, low density difference between particle and liquid and high viscosity of the liquid. As part of a multi-directional attack on the problem, ERDA has investigated basic principles in filtration of solids from the SYNTHOIL product. A high viscosity material containing 5.7 pct ash has been consistently batch filtered to below 0.1 pct, with some values as low as .035 pct. Several filter media were used, including filter paper and a precoat material that can be used on a continuous commercial filter. It was found that the nominal filter medium has little influence on the ultimate product clarity as the true filter medium is the cake that is formed during filtration. The nominal filter medium affects the pressure drop and time required to build a functioning cake. Filtration rates are very sensitive to temperature, but the relation is that of simple inverse proportionality to viscosity. A thin cake, readily distinguishable from the precoat, produces adequate clarity; additional cake increases pressure drop without substantial product improvement. This implies that a continuous precoat filter should be designed to maintain a thin coating of the product cake for optimum filtration.

I INTRODUCTION

All coal liquefaction processes must somewhere face the problem of separating unliquefied coal and the mineral matter that entered the process with the coal from the liquid stream that contains the product. One of the major advantages of the SYNTHOIL process now being developed by ERDA, as reported by Yavorsky *et al*. (1), is that it is quite tolerant of imperfections in the solids removal step, as both the product and the recycle stream are acceptable with as much as 10 pct undissolved material. However, a number of major advantages will be obtained when the present solids removal system is improved or replaced:

(1) Although SYNTHOIL ash values are low--in an experiment reported by Yavorsky *et al*. (2), a Kentucky coal containing 16.9 pct ash yielded a product oil with 1.5 pct, which represents excellent quality for a furnace designed to accept coal--reduction to below 0.1 pct will obviate the inclusion of electrostatic precipitators in the flue gas stream, thus reducing capital costs in product utilization.

(2) An ash content in the final product of less than .005 pct will meet specifications for gas turbine fuel, (3), thus greatly expanding the area of interchangeability between SYNTHOIL and scarce petroleum products.

(3) Reduction in the amount of organic insolubles will further reduce the sulfur content of the product, since the residual sulfur is concentrated in this component.

(4) Alternate systems, even if they should not improve the quality of the product, may yield substantial reductions in operating and capital costs.

Solids separation from coal liquefaction products is a difficult step because of the small particle size (4), variously estimated at mean diameters from 5 to below 0.5 micrometers, wide particle size distribution, from 25 micrometers down to colloidal size, low density difference between particle and liquid, physicochemical affinity between particle and liquid, and high liquid viscosity. We have examined the problem of improving product quality with centrifugation and with three approaches to filtration: (1) a bench filter consisting of a small pressure vessel with a perforated closure to support a 1.4-inch-diameter (0.0107 sq ft) filter medium of cloth, screen, filter paper or "precoat", a powder frequently used industrially as an easily renewable filter medium; (2) a tubular unit in which 0.7 sq ft of cloth filter medium in the form of a bag or sock is supported on a perforated steel tube so that the cloth can withstand high pressure differentials; and (3) a rotary drum pressure filter with 4.7 sq ft of filter surface, capable of operation as either a cloth filter or precoat filter. All three units are operable at 200° C and 200 psig.

Although there was some overlap, each unit was assigned different functions in the research. This paper describes the work with the bench filter. The bench filter was used to establish basic principles in SYNTHOIL filtration that

would be applicable to the engineering of the various commercial forms of filters. With the bench filter we have
shown that filtration is a feasible approach to ultra-low-ash
SYNTHOIL, that non-Newtonian behaviour of SYNTHOILS must
be considered in analyzing filtration, that the filter medium
which is initially installed in the filter controls the
initial flow rate and the length of time required to build
up a cake of the SYNTHOIL residue (during which time the
quality of the filtrate may be poor), after which flow rates
and product quality are relatively independent of the initial
medium and depend mostly on the quality and integrity of
the cake. Tests with a precoat medium showed that a sharp
demarcation could be obtained between the precoat and filter
cake, indicating that a continuous system could be operated
with good filtration and economic use of the precoat material.

II EXPERIMENTAL PROCEDURES

     The basic piece of equipment is shown in figure 1.
It consists of a perforated plate to support the filter
medium screwed to a steel case which can be pressurized
with nitrogen to drive the filtrate through the medium.
A top opening provides for insertion of a thermocouple.
The assembled unit can be inserted in a vertical 2-inch-
diameter tube furnace for operation at elevated temperature.
The experiments that were conducted in this equipment can
be subdivided into five groups listed in table 1. Groups
I, III and IV used a standardized feed (designated FB-36)

TABLE 1 *Experimental series in filtration*

Group	Table No.	Feed	Filter Material
I	2	FB-36	Filter paper 41F
II	3	Varied	Other sheet materials
III	4	FB-36	Precoat on paper
IV	5	FB-36, 38	Precoat on 100-mesh screen
V	–	Varied	Varied

which was a very viscous SYNTHOIL product, almost solid at
room temperature, that had been produced at 3,000 psi and
435° C from a Western Kentucky coal and contained 5.7 pct
ash and 1.0 pct sulfur. These high values of viscosity,
ash, and sulfur represent what we expect to be the most

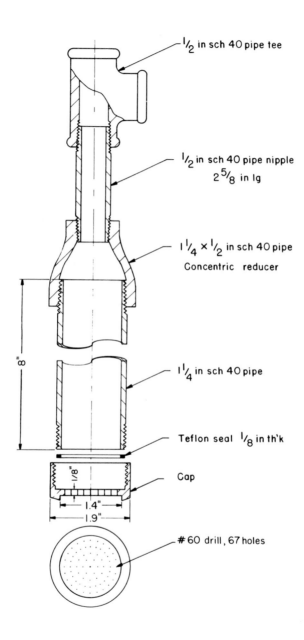

Fig. 1. *Small filter for use at elevated temperatures and pressures.*

difficult filtration parameters that would be met in the
SYNTHOIL process. Group II used other feeds in addition
to FB-36.

The first group of experiments, for which the data in
table 2 are typical, consisted of filtering FB-36 through
a standard laboratory filter paper (Whatman 41F)[1] at various
temperatures, pressures, and filter cake thicknesses, and
measuring flow rates and quality of product. In Group II,

*Table 2 Effect of pressure and temperature on SYNTHOIL
filtration through filter paper 41H*

==

Feed: FB-36, Ash 5.7%, Sulfur 1.0%

Run No.	07	08	10
Temperature, ° C	185	178	180
Pressure, psig	200	200	150
Filtrate, grams	17.5	20.0	20.0
Cake, grams	4.3	4.5	4.7
Initial[a] filtration rate, $lb\text{-}ft^{-2}\text{-}hr^{-1}$	4.72	4.91	2.40
Final[a] filtration rate ..	2.30	1.54	1.19
Ash, initial[a] filtrate, pct	0.21	0.27	0.18
Ash, final[a] filtrate, pct	.08	.06	.08
Sulfur, initial[a] filtrate, pct	.50	.50	.43
Sulfur, final[a] filtrate, pct	.50	.45	.44

[a] *"Initial" is first 10 grams, "final" is last ten
grams or less if 10 grams is not available.*

the Whatman 41H paper was replaced by other sheet filter materials. A major industrial filtration technique is the use
of rotary filters with "precoats" in which the filter medium
is a layer of granular material such as diatomaceous earth.
In the third group of experiments, two such materials were
used as a layer on top of filter paper, and in the fourth

[1] Appearance of commercial or trade names does not imply
endorsement of the products by the U. S. Government or
ERDA. Such names are used only for identification.

group, the precoat materials were used on top of an open
screen (100 mesh), without paper as a second medium. In both
these groups, the variables included coarseness of the precoat
material and the thickness of the precoat as well as those
variables measured in the previous groups. The fifth group
comprised experiments in which the feed materials were other
than FB-36. These data are not shown in a separate table, but
are incorporated into the other tables.

A.   Use of Filter Paper - Group I

The initial experiments on filter paper established
several facts. (1) For material as viscous as FB-36, temperatures greater than 165° C and pressures of 150 psig or more
were necessary to attain a filtration rate through the paper
that was experimentally practical in the sense of collecting
analyzable quantities of 10 to 20 grams in less than three
hours. (2) The flow rate is quite sensitive to temperature,
as shown in table 2 by comparison between tests 07 and 08
wherein a 7° C increase causes a 49 pct increase in "final"
flow rate. The comparison is made between "final" conditions
because, as will be verified later, "initial" conditions are
subject to many more rapidly changing and partially uncontrolled variables such as changing temperature and changing
resistance. Initial and final are indicated in quotes because,
although they are intended to represent point values of
changing variables, the nature of the measurements, as
indicated in the footnote of table 2, requires that they
are actually observations over substantial time increments.
(3) Flow rate is dependent on pressure. A 25 pct pressure
decrease between runs 08 and 10 produced a 23 pct "final"
flow rate decrease despite a 2° C temperature increase.
(4) Ash removal was at least 95 pct of the input ash, and
removal improved to 98.6 pct as the cake built up. (5)
The sulfur content on the other hand was rapidly reduced
to 50 pct of the input sulfur, but was not further reduced
with cake build-up.
The effects that have been described are indicative of
some of the differences between coal liquefaction products
and other materials that are usually subject to filtration.
The inability to reduce the sulfur content below some apparently fixed level implies that part of the sulfur in
SYNTHOIL is present in a colloidal or soluble component.
The temperature sensitivity of the flow rate is a reflection
of viscosity changes in the fluid. A series of viscosity
measurements were made using a rotational type of viscometer.

For a Newtonian or normal liquid, the shear stress *vs* rate-of-shear relation is a straight line passing through the origin, and the ratio of shear stress to rate of shear, which is the viscosity, is a constant that changes only with temperature. The relation demonstrated in figure 2,

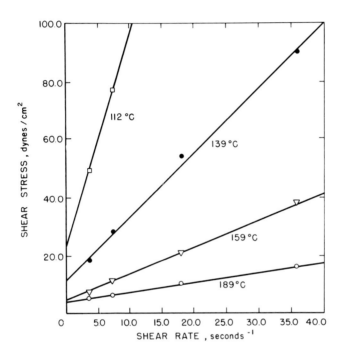

Fig. 2.   Viscous behavior of FB-36.

in which the curves do not pass through the origin, shows that the FB-36 feed material was non-Newtonian and that the effective viscosity is a function of flow rate and diameter of the flow channels (pores in a filtration medium) as well as temperature. This leads to a non-linearity of filtration rate with pressure drop when all other parameters are constant. The degree of non-linearity increases with lower flow rates, a factor which must be considered in all data extrapolation. Another complication in filtering a coal liquefaction material is that two insoluble components are actually being removed--the inorganic matter, which is measured as ash, and the insoluble organic matter, or incompletely converted coal. The two components have different particle size distributions, different affinities for the filter medium and different requirements with respect to

product quality specifications. Since the principle use
of the SYNTHOIL product is as a fuel, the incombustible
(inorganic) content of the filtrate is the measure of product
quality; however, the total material removed as cake (organic
plus inorganic) is the major factor controlling the separation process, as will be shown in the next section. Furthermore, the solubility of the organic matter in the feed may
be expected to increase with temperature, so that the temperature may affect not only viscosity but also the amount
of cake functioning as a filter medium in the filtration
process, and consequently will affect both filtration rate
and filtration quality.

B.   Effect of Filtration Septum - Group II

In the second group of experiments, three different
feed materials were filtered, each through two different
filtering sheet materials or septa. One septum of each
pair was the 41H Whatman filter paper used throughout the
Group I experiments. The others were of different construc-
tions and are described in table 3, which also lists the
results of the tests. In the table, the value for initial
filtrate rate is given as measured and also (in parentheses)
corrected to the value that would have been expected if
the flow rate were proportional to pressure and if the
initial pressure were 200 psi, which is the value for the
final period of all runs but 38. This calculation is necessary to make the values of initial and final filtration
rates more comparable. Experimentally, the initial pressures
had to be set much lower than the final values to keep the
initial flow rates from being too high to measure conveni-
ently.
In table 3, the most notable observation is that for the
run pairs 08/09 and 40/41 the initial flow rates and filtrate
clarity (pct ash) are highly disparate, whereas the final
conditions are quite similar. For example, tests 08 and
09, which were with the same feed material, had initial
corrected flow rates that differed by a factor of 168 whereas
the final rates differed by less than 2. Initial ash contents of the filtrates differed by a factor of 19 and final
values by 2. It may be hypothesized from this that initial
flow rates and clarity were controlled by the properties
of the septum that was used, but that the final filtering
medium was the same regardless of the septum and was essentially the filter cake itself. The influence of the septum,
therefore, was (a) to control the clarity of the initial
filtrate (which could be recycled later, if this clarity

TABLE 3 Effect of filter medium on SYNTHOIL filtration

Run No.	08	09	38	39	40	41
Pressure, psig	200	.1-200	.1-40	20-200	50-200	.2-200
Temp., °C	178	179	21	21	85	90
Feed	FB-36	FB-36	FB-39cf	FB-39cf	FB-39unc	FB-39unc
Viscosity @ temp., cp	85	85	151	151	8.4	6.8
Filter medium[a]	Wh41H	SFYC	Wh41H	Wh42	Wh41H	LC
Wt of filtrate, grams	20	22.4	132.4	142.3	93.3	67.2
Wt of cake, grams	4.5	4.0	0.11	0.16	15.0	12.2
Avg filtrate rate, lb-ft^{-2}-hr^{-1}	2.30	0.86	89.8	4.80	3.08	2.71
Initial[b] filtrate rate, (corrected to 200 psi) .	4.91(4.91)	4.13(826)	22.1(4420)	3.61(361)	6.60(26.4)	.87(868)
Final[b] filtrate rate	1.54	0.88	1,116.2	2.91	3.19	2.28
Pct ash, feed	5.7	5.7	0.012	0.012	7.4	8.9
Pct ash, initial[b] filtrate	0.27	5.1	.009	.008	0.01	1.9
Pct ash, final[b] filtrate .	.06	0.13	.010	.007	<.01	0.04

[a] Wh41H--Whatman filter paper, grade 41H (a rapid grade).
Wh42--Whatman filter paper, grade 42 (very retentive).
SFYC--Cloth, cotton non-woven (Eimco-Envirotech SFYC 1752116).
LC--A proprietary resin treated mat' (Lainyl Clartex, Lainiere de Selessin s.a., Belgium).
[b] "Initial" is first 10 grams, "final" is last 10 grams, or less if 10 grams is not available.

was below specification), (b) to control the time required
to build up a filter cake of sufficient thickness to become
an effective filtering medium. In short batches, filtered
without recycling, the septum will also affect the amount
of cake that is removed and formed into the filtering medium
and affect therefore both the average and final product
clarity.

In table 3, the observations for pairs 38/39 do not differ as much as those for the other two pairs, but this constitutes support for the filtering sequence suggested above.
Measurements of several filter cakes showed that 10 grams of
cake corresponded to a thickness of about 0.6 mm. Because of
the low solids content in the feed, the cake deposited during
runs 38 and 39 weighed only a fraction of a gram and was
therefore only a few micrometers thick which, based on later
estimates of particle size, would correspond to only a few
particle diameters and therefore provide a rather ineffective
filtering medium even at the end of the run.

C. Use of "precoat" as a Filter Medium - Groups III and IV

Industrial filtration is performed with many different
devices, but considerations of particle size, concentrations
and expected scale of operations in a coal liquefaction
industry suggest use of a rotary drum filter operated in
a "precoat" mode. In this mode, a layer of a granular
material, frequently a diatomaceous earth or a derivative
thereof, is deposited on a perforated rotating cylinder
which is partially immersed in the material to be filtered.
A pressure differential between the outside and inside of
the drum drives the liquid through the cylinder wall and
the solids are deposited on or within the layer of precoat.
As the drum rotates, the deposit is brought into contact
with a knife blade. The blade advances at a rate which
is set to remove the cake along with as much of the precoat
material as may be obstructed by penetration of the cake.
The next phase of the filtration program was an exploration of the parameters of "precoat" filtration *vis-a-
vis* rates and product clarity for other septa that have
been illustrated in tables 2 and 3. In the Group III experiments, as shown in table 4, two different grades of
a commercial calcined diatomaceous earth were deposited
from an oil suspension on grade 41H filter paper. Comparison
of the results for runs 07 and 08 in table 2 with those
for runs 15 and 11 in table 4, which pertained to the same
pressure, temperature and amount of cake, shows that the

TABLE 4 *Use of supported "precoat" as a filter medium*

Support: Wh-41-H filter paper

Run No.	11	12	13	14	16	17	15	27
Pressure, psig	200	200	200	200	200	200	200	200
Temperature, °C	170–179	164–171	160–179	180	182	183	182	185–178
Viscosity, cp ..[a].........	89–64	111–85	130–64	61	57	55	57	51–66
Precoat material ..[a]......	C545	C545	C545	C545	C545	C545	Hyflo	Hyflo
Precoat weight, grams	1.2	1.2	1.2	1.2	3.6	3.6	1.5	1.5
Feed	FB-36	FB-36	FB-36	FB-36	FB-36	FB-36	FB-36	FB-36
Filtrate, grams	21.1	43.8	21.8	40.9	22.9	46.8	22.2	50.5
Cake, grams	4.7	9.5	4.7	8.8	6.4	9.4	5.4	10.0
Initial[b] filtrate rate, corrected to final press., lb-ft^{-2}-hr^{-1}	5.51	24.9	13.7	13.9	25.6	60.5	12.1	38.4
Final[b] filtrate rate	1.83	0.91	2.80	1.07	9.02	2.19	1.82	1.02
Pct ash, initial[b] filtrate.	0.23	0.38	0.43	0.83	0.57	0.07	0.12	0.11
Pct ash, final[b] filtrate ..	.03	.07	.08	.22	.05	.01	.08	.04

[a] Precoat: C545 – Johns–Manville Celite 545, Nominal relative flow rate: 2,160
    Hyflo – " " Hyflo Super-Cel, Nominal relative flow rate: 500
[b] See footnote, table 2.

superposition of precoat on filter paper caused the initial
flow to be increased considerably and the initial ash to
be decreased; the final flow rates, after proper allowance
for point values of temperature and viscosity differences,
were the same within limits of experimental error. Final
ash removal with precoat was equal to or somewhat better
than with no precoat. The observation that the presence
or absence of precoat caused substantial differences at
the beginning of the run but small differences at the end
(by which time a cake had built up) provides confirmation
of the hypothesis advanced in the previous section that
the effective filtration medium is not the original septum
but the layer of cake that is filtered from the feed.
The presence of the precoat diffuses the solids during the
initial separation so that, during this period, formation
of a dense cake is delayed, but ultimately a cake is built
up, pressure drop increases, and product clarity improves.

Table 4 also permits comparisons between (a) grades
of precoat, (b) different amounts of the same grade, (c)
different amounts of feed and therefore of cake, and (d)
different temperatures. The table shows no significant
correlation between any of these variables and the quality
of the product, which is again confirmation of the hypothesis
that the ash removal is primarily a function of the deposited
cake rather than of the substrate. Run 14 was included
in table 4 to illustrate the affect of some of the variables
on flow rates. However, product quality in run 14, as indicated by ash analysis, is obviously inconsistent with that
for the other runs in the table. The later discussion in
connection with table 5 will illustrate how by-passing of the
filter medium by an amount that would be negligible with
respect to flow rate can produce major changes in ash content
of the product.

The fact that the amount of cake has no obvious affect
on quality (compare, for example, pairs 11/12 and 15/27)
implies that, once the cake is formed, the solids separation
is performed at the surface of the cake rather than within
its depth. A consistent observation is that, for each pair
of runs that is identical except for amount of charge (11/12,
13/14, etc.), the final filtration rate is less for the
greater charge. That is, the approximately doubled thickness
of cake causes the flow rate to be cut approximately in
half. This observation, in conjunction with the observation
that the extra cake does not contribute to improved product
clarity, leads to the conclusion that a continuous system
should be operated with the minimum cake thickness required

to give good clarity. Since, however, a finite cake layer
is required for acceptable clarity, the cake-on-precoat
layer should not be cut so deep as to remove the entire
cake layer, thus achieving the additional advantage of
preserving all the precoat.

"Initial" flow rates are less meaningful than subsequent
measurements because they are not instantaneous values but
are values integrated over onehalf to one-third the total
filtration time during that interval when pressures, temperatures (and therefore viscosities), and cake formation
are all changing. In all cases, the "initial" flow rate
for the larger amount of feed was greater than for the
smaller amount; this is probably not of fundamental significance but is a reflection of the longer interval to reach
equilibrium and perhaps indicates that the larger volumes
may initially disturb the upper layer of the precoat material.

The effect of the larger amount of precoat (compare
run 16 with 13, or run 17 with 14) is to cause a larger
"initial" flow rate, which supports the hypothesis offered
in an earlier paragraph that the porosity of the precoat
delays the formation of a filtering cake. However, Fyflo
(the less porous precoat as rated by the manufacturer) used
in run 27 increased the flow rate over that measured in
run 14, in which Celite 545, nominally a more porous precoat,
was used. The difference in this case may be due to the
fact that Fyflo and Celite 545 are somewhat different
chemically, so that the flow resistance to water, as used
in the manufacturer's criterion, may not follow the same
patterns as resistance to SYNTHOIL, if hydration, streaming
potentials or other physicochemical forces are operative.

The last series of filtrations was made with the precoat
material supported only on a 100-mesh stainless steel screen,
i.e., unbacked by filter paper. Initial tests gave erratic
results that could be explained by the sensitivity of the
precoat layer to small disturbances in flows and pressures.
Even the minor fluctuations in pressure caused by normal
operation of a pressure regulator created occasional rapid
flow which disturbed the precoat layer and caused the feed
to bypass the precoat, as disclosed by disassembling the
filter unit and observing layers of SYNTHOIL between the
precoat and the wall of the filter tube. Successful filtrations were obtained by installing a needle valve between
the pressure regulator and the filter to damp out sudden
fluctuations and by maintaining close control on temperature

changes. The data in table 5 were taken for five runs
permitting comparison between two precoat materials and
two different feed materials.

Because, for the unsupported precoat runs, it was deemed
necessary to increase temperatures and pressures very gradu-
ally, the conditions for the "initial" interval varied widely
from run to run and these values are less comparable than
are the "final" data. In all cases of good filtration,
a line of sharp demarcation was found between the cake and
precoat, which implies that little penetration takes place
into the precoat and that, on a continuous filter, the cake
could be removed without excessive loss of precoat.

The data in table 5 confirm the previously stated
hypothesis that once a stable filter cake layer is formed,
SYNTHOIL filtration is independent of the substrate on which
the layer is deposited. All the "final ash" analyses represent removal of 98.8 to 99.3 pct of the ash in the feed.
The apparently wide variation in the absolute value of the
ash analysis (.08 pct to .035 pct) reflects a sensitivity
of the process to bypassing: if the filter actually reduces
the ash content from 5.7 pct to .035 pct, and only 1.0 pct
of the feed bypasses the filter action, the product will
contain (.035 x 0.99) + (5.7 x 0.01) = .092 pct ash, an
increase of 2.5 times the true filtrate analysis.

The qualitative effect of temperature is apparent in
table 5. Runs 32, 33, and 34 differ only in temperature,
and the observed filtration rates for the three experiments
increase as the viscosities decrease. Additional data is
available for an analysis of the pressure drop-flow relation,
but since no direct measurements of the cake deposited are
available except at the end of the run, and since flow rates
were not measured but only total weights at various times,
a number of assumptions must be made and only a simplified
correlation can be attempted. We may assume that:

(a) the flow rate $dW/d\Theta$ is proportional to the driving
    force ΔP, inversely proportional to the fluid pro-
    perties incorporated in the measured viscosity η,
    and inversely proportional to a cake property we
    define as a resistance R. This assumption may be
    restated as

$$dW/d\Theta = k_1 \Delta P/(\eta R) \qquad\qquad 1)$$

Table 5 Use of unsupported "precoat" as a filter medium

Run No.	29	32	33	34	37
Pressure, psig	100-150	28-200	30-100	30-100	15-50
Temperature, ° C	152-166	155-179	170-179	190-200	96
Viscosity, cp ...;^{a}.....	179-103	159-64	89-64	43-31	63
Precoat, material	Hyflo	C545	C545	C545	C545
Precoat weight, grams ...	6.0	6.0	6.0	6.0	6.0
Feed	FB-36	FB-36	FB-36	FB-36	dFB-38
Filtrate, grams	19.2	40.3	42.8	45.4	60.2
Cake, grams^b	9.4	(7.4)	9.2	(10.0)	9.2
Initial^c filtrate rate, corrected to 200 psig, lb-ft^{-2}-hr^{-1}	4.72	11.0	28.3	47.7	72.1
Final^c filtrate rate, corrected to 200 psig, lb-ft^{-2}-hr^{-1}	2.87	1.35	1.84	2.65	114.0
Ash, initial^c, pct	0.16	0.73	0.10	0.12	0.03
Ash, final^c, pct	.04	.056	.067	.035	.08
Sulfur, initial^c, pct ...	.51	.70	.52	.41	.24
Sulfur, final^c, pct	.51	.53	.50	.47	.26

^{a} See footnote 1, table 4.
^{b} Cake weight in parentheses denotes indirect determination from measurement of cake thickness.
^{c} See footnote 2, table 2.
^{d} Ash = 8.6 pct, Sulfur = .37 pct.

which may be used as a defining equation for R
by setting $k_1 = 1$, from which

$$R = \Delta P/(\eta \; dW/d\Theta) \qquad \qquad 2)$$

As it is reasonable to assume that:

    (b) the resistance to flow will be inversely propor-
tional to the cross-sectional area transverse to
the direction of flow, Equation 2 can be somewhat
generalized by stating the flow rate as flow per
square foot or:

$$R = \Delta P/[\eta(dW/d\Theta)/A] \qquad \qquad 3)$$

The test procedures that were followed provide no mechanism
for determining the amount of cake present at any instant
during the filtration; we therefore make a third assumption:

    (c) the amount of cake is proportional to the amount
of slurry that has been filtered.

Figure 3 is a graph of the resistance to flow through
the filter cake and precoat of runs 32, 33, and 34, plotted
as a function of the total weight of slurry filtered.
Although the legend in the figure indicates an average
temperature for each run, the resistance values for each
point were calculated from the measured temperature at the
specific point.   Figure 4 is a similar plot for earlier runs
in which filter paper and not precoat was used. It is apparent that the resistance, as defined in terms of pressure drop
and flow, and as corrected for temperature effects by division
by the viscosity, correlates well with the total weight of
filtrate. If the abscissa were weight of cake and all
other cake properties were invariant, the curves of figures
3 and 4 would be straight lines, with the y intercept equal
to the resistance of the initial septum, i.e., the precoat
or filter paper. The observed curvature could be due to
any one or more of a number of factors which have been
mentioned earlier.

    (a) The weight of cake is not proportional to the weight
of filtrate. This is, in fact, known to be the
case, as the analytical results showed that more
solids were removed from the last part of the
filtrate than from the "initial" filtrate. However, changes in the weight of ash removed from
the "final" filtrate are small so that the last

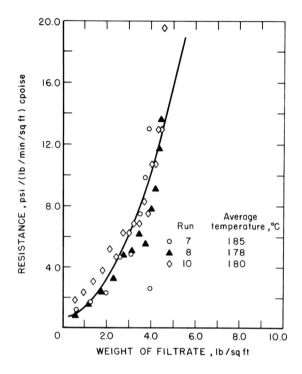

*Fig. 3. Filtration resistance through filter cake
and 6 grams of C-545 precoat.*

half of the data should approximate a straight
line.

(b) Size distribution of the particles removed may
change, so that more fines are incorporated into
the cake that is formed during the later part of
the run.

(c) The cake may be compressible, so that its permea-
bility is decreased as the pressure drop across it
increases.

(d) The non-Newtonian viscous behavior of the fluid,
as demonstrated in figure 2, may be of sufficient
magnitude that the changed shear rate in the latter
part of the filtration may create a much greater
effective viscosity than is used in the resistance
calculation.

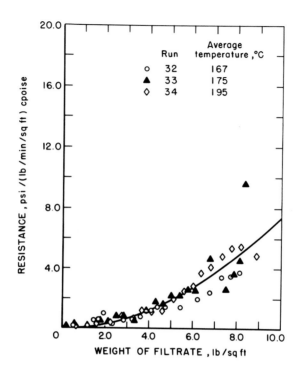

Fig. 4. Filtration resistance through filter cake and filter paper.

Each of the above possibilities presents a major challenge, the solution of which may lead to greatly improved filtration. However, even without resolving the reasons for the curvature of figures 3 and 4, important deductions can be made.

(a) Resistance increases more than proportionally to the amount of cake. For a continuous system where the cake thickness can be controlled, the thickness should be held at the minimum value that will yield acceptable filtrate clarity.

(b) Resistance per unit of cake increases much more rapidly when the initial septum is filter paper rather than precoat.

D. Varied Feeds - Group V

Although most of the tests discussed were made with
the feed material designated as FB-36, several other materials were used. The results for these materials were
entirely consistent with the results for FB-36 and they
have been included in tables 3 and 5 (runs 37, 38, 39, 40,
41) wherein the only significant effect due to the feed
is that feeds with lower ash content yielded products with
less ash, although, for reasons discussed earlier in connection with table 3, the changes were not proportionate.

III ADDITIONAL WORK

In parallel efforts, filtration studies have been
carried on with larger scale equipment. In tests with a
rotary drum filter, details of which will be published at
a later date, we have utilized the principles advanced here
of maintaining a thin, uniform cake of solids on top of
a precoat layer. A recent run of 444 uninterrupted hours
on stream included a steady state period of 200 hours during
which an ash removal of at least 99.1 pct was maintained,
with an average filtrate content of .09 pct ash.

IV CONCLUSIONS

Batch filtration is a technically feasible method of
removing ash from a SYNTHOIL product, so that the only
remaining problem is the mechanics of scaled-up continuous
filters. Starting with a material containing 5.7 pct ash,
analyses of less than 0.1 pct have been almost routinely
attained using a commercial precoat material with no filter
paper support. The nominal filter medium has little influence on the ultimate product clarity, as the true filter
medium is the cake that is formed from the feed. The nominal
filter medium affects the pressure drop, the time required
to build up a functioning cake, and the number of precautions
that are required to forestall possible system disturbances
such as pressure variations during cake build-up. Filtration
rates are very sensitive to temperature, but when cake
resistances are divided by the measured viscosity, the
temperature effect is properly accounted. In the batch
filter tests reported here, two observations suggest that
a continuous filter will be effective:

(1) a clear demarcation usually exists between cake and precoat so that separation in a continuous unit should be simple with minimum cost in precoat;

(2) adequate clarity is attained with a relatively thin cake layer whereas additional cake adds to the pressure requirement without major improvement in clarity.

Filtration of SYNTHOIL causes a certain amount of reduction in sulfur, but beyond this point improved filtration, as measured by ash reduction, does not improve sulfur reduction. Quantitative analysis and possible improvement of the filtration process requires further study into particle size distribution, cake compressibility, and the effect of the observed non-Newtonian flow characteristics of the fluid being treated.

V   NOMENCLATURE

W = weight of filtrate
Θ = elapsed time
ΔP = pressure difference across filter
η = viscosity
R = resistance to filtrate flow
k_1 = constant
A = area of filter normal to flow

VI   REFERENCES

1. Yavorsky, P. M., Akhtar, S., Friedman, S., Chem. Eng. Prog., 69, No. 3, 51-52 (1973).

2. Yavorsky, P. M., Akhtar, S., Friedman, S., AIChE Symposium Series No. 137, Vol. 70, 101-105 (1974).

3. General Electric Co. "Gas Turbine Liquid Fuel Specifications" GEI-41047E, March, 1973, page 5 (1973).

4. Walker, P. L., Electric Power Research Institute, Report No. EPRI-366-1, December, 1975.

ANALYZING LIQUID PRODUCTS DERIVED FROM COAL CONVERSION PROCESSES

J. E. Dooley and C. J. Thompson
Bartlesville Energy Research Center

Coal derived liquids were characterized by means of separation
techniques such as distillation, gradient elution chromatography,
acid and base extractions, and gel permeation chromatography
(GPC) followed by instrumental analyses such as mass spectrometry and nuclear magnetic resonance spectrometry. Samples
from the COED and SYNTHOIL liquefaction processes were
studied. Results show liquids are amenable to the characterization procedure and provide basic data that should be useful in
formulating refining processes for these materials. The procedure
provides considerable detail in the analysis and relates chain
carbons, naphthenic ring carbons, and aromatic ring carbons
to established GPC-mass spectral correlations.

I. INTRODUCTION

Major coal-conversion processes under consideration (1) for
development by U.S. interests are capable of producing a wide variety
of products. Processing and using these materials in future applications will require a better understanding of their composition than has
been necessary with conventional energy sources. To provide basic
data, the Bartlesville Energy Research Center has been studying liquid
products derived from various coal-conversion processes.
Depending upon the material to be processed and the information
desired, numerous approaches to appropriate characterization may be
taken. Some investigators have used mass spectrometry as a single
technique to analyze materials of wide boiling range without prior
separations (2, 3). These mass spectral methods are well conceived
and furnish useful information, but cannot provide the compositional

detail that is possible with prior separations. Further, more accurate type assignment and quantitation are possible with preliminary separations. Limited analytical data that provide physical characteristics (4) such as percent oil, asphaltenes, gravity, viscosity, etc. may be satisfactory for monitoring coal liquefaction process operations and for comparing materials that will satisfy utility type fuel requirements. However, the composition of coal liquids to be further upgraded such as those to be used for transportation, needs to be determined in more detail to provide the refiner with the necessary basis for processing these materials.

The Bartlesville Energy Research Center is using the approach of physical and chemical separation, followed by instrumental characterization, to provide considerable detailed data on the composition of coal liquefaction products. The key to the identification procedure for those materials boiling above 200° C lies in the GPC-mass spectral correlation method developed at Bartlesville (5, 6). Most of the procedures used were developed for the analysis of high-boiling petroleum fractions during a cooperative project between the U.S. Bureau of Mines and the American Petroleum Institute (7). Some minor modifications have been necessary to adapt the procedures to the products of coal liquefaction.

Materials from three different coals and two liquefaction processes were studied: two COED materials produced from Utah and western Kentucky coals and a third material from the Synthoil process using West Virginia coal. The COED process involves pyrolysis of the coal followed by catalytic hydrogenation of the crude liquid, whereas the Synthoil process is a direct catalytic hydrogenation of coal slurried in a recycle liquid. Although process conditions are not available for the two COED products, the Synthoil product was prepared by processing the coal at 450° C and 4,000 psig (8). Hydrogen and a slurry of 35 percent coal in recycle oil were fed at 25 pounds per hour into a nominally half-ton (slurry) per-day unit with a 14.5-foot-long catalytic reactor. The catalyst was 1/8-inch pellets of cobalt molybdate on silica alumina. Two other liquid products from the H-Coal process are under investigation, and will be completed soon.

II. EXPERIMENTAL

The general procedure used in the study of full-boiling-range coal liquid products is shown in Figure 1. Boiling ranges of distillates were

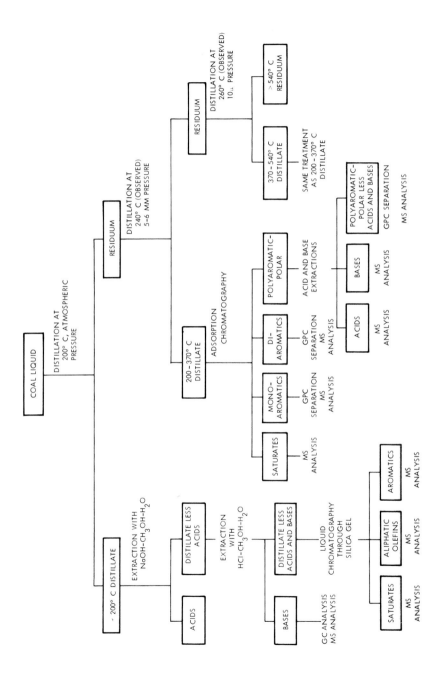

Fig. 1. Procedure for characterizing coal liquids.

were checked by simulated distillation (9) to establish the approximate boiling ranges desired. In previous work with petroleum crudes, care was exercised to prepare distillates having identical boiling ranges so that more precise comparisons of data could be made. However, smaller samples of the available coal liquids have precluded adjustment of still conditions to any great extent, and the boiling ranges as determined by simulated distillation for the distillates prepared are not exactly the same. This was not considered essential for these initial studies because the primary interest was to determine whether the procedure might be effective and the adjustments necessary to make an effective characterization. All distillates were prepared in a 4-inch-diameter Rota-Film molecular still. The still, a continuous-flow, wiped-wall vessel, provides minimum residence time of the sample at elevated temperatures, thus limiting thermal degradation of the material being processed. Material is passed through the still for each set of conditions of temperature and pressure at the rate of 600 to 1,000 ml/hr.

After the distillates were prepared, the material boiling below about 200° C was separated further into acids, bases, and hydrocarbon-neutral fractions for analysis by gas chromatography (GC), ultraviolet (UV) fluorescence, and mass spectrometry. Distillates boiling from about 200° C to 370° C and 370° C to 540° C were processed through adsorption columns (10) to produce four concentrates for each distillate: saturates, monoaromatics, diaromatics, and polyaromatic-polar material. The saturates were analyzed directly by mass spectrometry; monoaromatics and diaromatics were separated by gel permeation chromatography (GPC) and analyzed by GPC-mass spectral correlations (5, 6); and the polyaromatic-polar concentrate was separated into acids, bases, and hydrocarbon-neutrals. The polyaromatic-polar concentrate less acids and bases was then separated by GPC and characterized in the same manner as the monoaromatics and diaromatics.

The adsorption columns used were about 8 feet long by 1 inch diameter, packed with 28 to 200 mesh, Davison grade 12 silica gel in the top half and 80 to 200 mesh Alcoa F-20 alumina in the bottom half, and were operated downflow. The GPC column was about 16 feet long by 1 inch diameter, packed with 100 Å polystyrene gel in the top half and 400 Å polystyrene gel in the bottom half, and was operated downflow. The mass spectrometer used for these studies was a CEC 21-103C low-resolution instrument; saturates were analyzed by high-ionizing voltage spectra and aromatics by low-ionizing voltage spectra. High-resolution mass spectrometers, the AEI MS-30 and CEC 110, were also used to resolve certain heteroatomic species that overlapped the same nominal hydrocarbon series. NMR spectra from a Varian A-60 instrument were obtained to determine the proton distributions across the GPC runs and thereby confirm and enhance the findings of the GPC-mass spectral correlations.

III. RESULTS AND DISCUSSION

Summary data shown in Tables I and II were selected from comprehensive characterization studies (11 - 13) that have been completed for each coal liquid. Table IA summarizes distillate distributions, Table IB shows totals for major compound classes, and Table II lists ring number distributions by compound class for each distillate. As shown by the distillate distributions data in Table IA, the Synthoil material from West Virginia coal is the highest boiling of the three, having 95.6 percent boiling above about 200° C and 25.7 percent residuum (boiling above about 540° C). Utah syncrude is rated second in higher-boiling material with 85.7 percent boiling above about 200° C. Western Kentucky shows 78.4 percent boiling above 200° C.

Table IB shows some indication of the degree of aromaticity for each syncrude. For example, assuming the 25.7 percent residuum shown for Synthoil is mostly aromatic, Synthoil then would be the most aromatic of the three having a total aromatic hydrocarbon content near 70 percent, whereas western Kentucky and Utah syncrudes are about 54 percent and 49 percent, respectively.

The quantities of sulfur and nitrogen determined for each coal liquid shown in Table IA and the amounts of acids, bases, and saturates listed in Table IB reflect, to some extent, the degree of hydrogenation for each product. For example, the data suggest that the western Kentucky COED product probably was more severely hydrogenated than the Utah COED product, although coal structure or other processing conditions could account for some of the differences in composition. No direct comparison of the Synthoil product with the COED products could be made because of the substantial differences in processing as well as the coal source.

Further insight can be gained into the cyclics present in the hydrocarbon structures of the three coal liquids by examination of the data in Table II. From the data on saturates, note the lower concentrations of total paraffins as compared to total cyclics. For the Utah syncrude,

TABLE 1
Distillates and Major Compound Types Separated From Three Coal
Liquids
==

	COED Syncrude Utah Coal	COED Syncrude Western Kentucky Coal	Synthoil West Virginia Coal
	Wt. Pct.	Wt. Pct.	Wt. Pct.
A. COAL LIQUIDS AND DISTILLATES			
COAL LIQUIDS:			
Sulfur	0.05	0.08	0.42
Nitrogen	0.48	0.23	0.79
DISTILLATE DISTRIBUTIONS:			
< 200° C Distillate[1]	13.3	21.0	4.4
200–370° C Distillate[2]	45.4	54.2	42.6
370–540° C Distillate[3]	40.3	24.2	27.3
540° C+ Residuum[4]	–	–	25.7
Losses	1.0	0.6	–
B. MAJOR COMPOUND TYPE DISTRIBUTIONS			
Total Saturates	30.10	33.42	10.65
Total Monoaromatics	20.06	33.64	14.23
Total Diaromatics	14.41	12.77	15.53
Total Polyaromatics	14.48	7.91	14.61
Heteroatomic Species	1.50	0.29	2.24
Acids	8.17	3.11	10.40
Bases	1.30	1.05	2.75
Residuum (Not Analyzed)	None	None	25.7
Other Material Not Analyzed	9.98	7.81	3.89
Totals	100.00	100.00	100.00

[1] < 204° C for Utah, < 205° C for western Kentucky, and < 207° C for Synthoil.

[2] 204–381° C for Utah, 205–380° C for western Kentucky, and 207–363° C for Synthoil.

[3] 381° C+ residuum for Utah, 380° C+ residuum for western Kentucky, and 363–531° C for Synthoil.

[4] Utah and western Kentucky syncrudes had no material boiling above about 540° C.

TABLE 2
Ring Number Distributions of Hydrocarbon Concentrates From Three Liquids

Concentrate and total number of rings determined (includes Aromatic and Aliphatic rings)	Syncrude From COED Process Utah Coal			Syncrude From COED Process Western Kentucky Coal			Synthoil Liquid Product From West Virginia Coal		
	<204°C Distillate	204-381°C Distillate	381°C+ Residuum	<205°C Distillate	205-380°C Distillate	380°C+ Residuum	<207°C Distillate	207-363°C Distillate	363-531°C Distillate
Weight Percent of Total Liquid Product									
SATURATES:									
Paraffin	–	4.06	4.67	–	2.79	1.47	–	1.09	0.78
1-ring	–	1.39	1.09	–	3.34	1.00	–	1.71	0.91
2-ring	–	1.65	0.50	–	3.22	0.67	–	1.59	0.17
3-ring	–	1.11	1.09	–	2.41	0.75	–	1.52	0.23
4-ring	–	1.58	0.67	–	1.37	0.69	–	0.89	0.26
5-ring	–	2.85	1.11	–	0.40	0.53	–	–	0.31
6-ring	–	–	1.30	–	–	0.44	–	–	–
TOTALS	7.05	12.64	10.41	14.34	13.53	5.55	1.19	6.80	2.66
MONOAROMATICS:									
1-ring	–	1.09	0.66	–	1.76	0.28	–	0.21	0.09
2-ring	–	4.50	0.60	–	7.83	0.39	–	0.90	0.05
3-ring	–	3.91	0.88	–	8.43	0.80	–	4.75	0.09
4-ring	–	1.52	1.60	–	3.56	2.11	–	4.62	0.23
5-ring	–	0.41	1.33	–	1.04	1.38	–	1.15	0.42
6-ring	–	0.01	0.55	–	0.19	0.69	–	0.02	0.35
7-ring	–	–	0.16	–	–	0.32	–	–	0.10
8-ring	–	–	0.05	–	–	0.11	–	–	0.04
TOTALS	2.79	11.44	5.83	4.75	22.81	6.08	1.21	11.65	1.37

TABLE 2
Ring Number Distributions of Hydrocarbon Concentrates From Three Liquids -- Continued

Weight Percent of Total Liquid Product

Concentrate and total number of rings determined (includes Aromatic and Aliphatic rings)	Syncrude From COED Process Utah Coal			Syncrude From COED Process Western Kentucky Coal			Synthoil Liquid Product From West Virginia Coal		
	<204°C Distillate	204-381°C Distillate	381°C+ Residuum	<205°C Distillate	205-380°C Distillate	380°C+ Residuum	<207°C Distillate	207-363°C Distillate	363-531°C Distillate
DIAROMATICS:									
2-ring	-	3.82	0.31	-	2.06	0.08	-	3.21	0.05
3-ring	-	3.41	0.91	-	3.11	0.72	-	3.98	0.64
4-ring	-	1.11	1.80	-	1.46	1.69	-	1.68	1.51
5-ring	-	0.28	1.53	-	0.31	1.63	-	0.34	1.22
6-ring	-	-	0.77	-	0.06	0.98	-	0.01	1.17
7-ring	-	-	0.34	-	0.01	0.47	-	-	0.99
8-ring	-	-	0.13	-	tr	0.16	-	-	0.36
9-ring	-	-	-	-	-	0.04	-	-	0.16
10-ring	-	-	-	-	-	-	-	-	0.05
11-ring	-	-	-	-	-	-	-	-	0.02
TOTALS	-	8.62	5.79	-	7.00	5.77	0.14	9.22	6.17
POLYAROMATICS:									
3-ring	-	0.80	1.14	-	1.11	0.37	-	0.83	0.20
4-ring	-	1.29	2.36	-	1.07	1.04	-	1.48	1.45
5-ring	-	0.75	2.88	-	0.42	1.20	-	0.79	1.42
6-ring	-	0.26	3.32	-	0.21	1.04	-	0.15	2.03
7-ring	-	0.11	1.56	-	0.09	1.15	-	0.08	2.20
8-ring	-	-	0.01	-	0.02	0.16	-	0.03	2.01
9-ring	-	-	-	-	-	0.03	-	0.02	1.03
10-ring	-	-	-	-	-	-	-	0.01	0.69
11-ring	-	-	-	-	-	-	-	0.01	0.18
TOTALS	-	3.21	11.27	-	2.92	4.99	-	3.40	11.21

total paraffins in the saturates amount to about 8.7 percent of the syncrude and cyclics about 14.3 percent, which means the total saturates
are about 63 percent cyclic. Western Kentucky syncrude saturates
show about 4.3 percent of the syncrude as total paraffins and 14.8 percent as cyclics, or about 78 percent of the saturates are cyclic. The
Synthoil product saturates show a paraffin content of 1.9 percent of the
total liquid and cyclics content of 7.6 percent, giving an approximate
value of 80 percent cyclics in the saturate concentrates. Since Utah
syncrude and western Kentucky syncrude were produced by the same
process, the significant difference in the amount of paraffins produced
would indicate some possible differences in the structures of the coal
sources, although the lack of details on the processing of both coals
makes this observation tentative from the distributions shown in Table
2, the aromatics which contain four or more total rings add up to about
23.1, 22.6, and 14.4 percent of the Utah syncrude, western Kentucky
syncrude, and Synthoil product, respectively. These data would indicate that the Utah and western Kentucky syncrudes may contain more
of the larger aromatic ring systems than the Synthoil product; however,
as shown in Table 1B, 25.7 percent of the Synthoil product is residuum
that was not analyzed (and probably is composed of multi-ring systems),
and about 13.2 percent of the product is in acids and bases. Total
ring number distributions shown for each major compound class indicate
the two-, three-, four-, and five-ring systems to be predominant in all
three syncrudes, Synthoil having material up to 11 total rings. Mass
spectral and GPC correlation data and NMR data for these materials
indicate structures somewhat more condensed, with more short alkyl
groups attached to the cyclic nucleus, than those found in similar boiling ranges of petroleum (14 – 18). We expect to establish soon the
amount of condensation and structural arrangement of materials in the
GPC fractions to provide more useful information on the cyclic structures for determining proper refining processes.

IV. CONCLUSIONS

In general, the scheme followed in separation of syncrude materials
for characterization studies has provided data for determining appropriate refining processes for these materials. The preliminary separations
provide more meaningful fractions for accurate analysis by mass spectrometry and other instrumental techniques by type identification and
quantification. The refiner should find this type of data useful in the
selection and development of processes for upgrading coal liquids to
finished products.

For the three materials examined, Synthoil product appears to be the most aromatic, with ring systems having up to about 11 total rings. Differences between the two COED syncrudes may be attributable to the severity of hydrogenation of the crude pyrolysis liquids, although insufficient information is available to distinguish between the effects of process conditions and character of coal used as raw material. In continued studies, the characterization and analysis of additional liquids from known combinations of coal source, liquefaction process, and degree of upgrading will provide a more meaningful basis for future refining processes. For more meaningful data, future studies will require access to information such as the coal source and process conditions.

V. REFERENCES

1. "Coal Conversion Activities Picking Up,"pp. 24-25. Chem. & Eng. News, Dec. 1, 1975.
2. Aczel, T., Foster, J. Q., and Karchmer, J. H., "Characterization of Coal Liquefaction Products by High Resolution-Low-Voltage Mass Spectrometry," Vol. 13, No. 1, pp. 8-17. Am. Chem. Soc., Div. Fuel Chem., Prepr., April 1969.
3. Swanziger, J. T., Dickson, F. E., and Best, H. T. "Liquid Coal Compositional Analysis by Mass Spectrometry," Vol. 46, No. 6, pp. 730-734. Anal. Chem., May 1974.
4. Sternberg, Heinz W. "The Nature of Coal Liquefaction Products," Symposium on Progress in Processing Synthetic Crudes and Resids. Presented before the Div. of Pet. Chem., Am. Chem. Soc., August 24-29, 1975.
5. Hirsch, D. E., Dooley, J. E., and Coleman, H. J. "Correlations of Basic Gel Permeation Chromatography Data and Their Applications to High-Boiling Petroleum Fractions," 77 pp. US Bur of Mines, Rep. of Invest. 7875, 1974.
6. Hirsch, D. E., Dooley, J. E., Coleman, H. J., and Thompson, C. J. "Qualitative Characterization of 370° to 535° C Aromatic Concentrates of Crude Oils From GPC Analyses," 26 pp. US Bur of Mines, Rep. of Invest. 7974, 1974.
7. Haines, W. E., and Thompson, C. J. "Separating and Characterizing High-Boiling Petroleum Distillates: The USBM-API Procedure," 30 pp. ERDA-LERC Rep. of Invest. 75/5, ERDA-BERC Rept. of Invest. 75/2, July 1975.
8. Friedman, Sam, Yavorsky, Paul M, and Akhtar, Sayeed. "The Synthoil Process," pp. 481-494. Clean Fuels from Coal,

Symposium II Papers, Inst. of Gas Technol., June 1975.
9. Coleman, H. J., Dooley, J. E., Hirsch, D. E., and Thompson, C. J. "Compositional Studies of a High-Boiling 370-535° C Distillate from Prudhoe Bay, Alaska, Crude Oil," Vol. 45, No. 6, pp. 1724-1734. Anal. Chem., August 1973.
10. Hirsch, D. E., Hopkins, R. L., Coleman, H. J., Cotton, F. O., and Thompson, C. J. "Separation of High-Boiling Petroleum Distillates Using Gradient Elution Through Dual-Packed (Silica Gel-Alumina Gel) Adsorption Columns," Vol. 44, No. 6, pp. 915-919. Anal. Chem., May 1972.
11. Dooley, J. E., Sturm, G. P., Jr., Woodward, P. W., Vogh, J. W., and Thompson, C. J. "Analyzing Syncrude From Utah Coal," 24 pp. ERDA-BERC Rep. of Invest. 75/7, August 1975.
12. Sturm, G. P., Jr., Woodward, P. W., Vogh, J. W., Holmes, S. A., and Dooley, J. E. "Analyzing Syncrude From Western Kentucky Coal," 27 pp. ERDA-BERC Rep. of Invest. 75/12, November 1975.
13. Woodward, P. W., Sturm, G. P., Jr., Vogh, J. W., Holmes, S. A., and Dooley, J. E. "Compositional Analyses of Synthoil From West Virginia Coal," 22 pp. ERDA-BERC Rep. of Invest. 76/2, January 1976.
14. Dooley, J. E., Hirsch, D. E., Thompson, C. J., and Ward, C. C. "Analyzing Heavy Ends of Crude (Comparisons of Heavy Distillates)," Vol. 53, No. 11, pp. 187-194. Hydrocarbon Process., November 1974.
15. Thompson, C. J., Dooley, J. E., Hirsch, D. E., and Ward, C. C. "Analyzing Heavy Ends of Crude (Gach Saran)," Vol. 52, No. 9, pp. 123-130. Hydrocarbon Process., September 1973.
16. Dooley, J. E., Thompson, C. J., Hirsch, D. E., and Ward, C. C. "Analyzing Heavy Ends of Crude (Swan Hills)," Vol. 53, No. 4, pp. 93-100. Hydrocarbon Process., April 1974.
17. Dooley, J. E., Hirsch, D. E., and Thompson, C. J. "Analyzing Heavy Ends of Crude (Wilmington), Vol. 53, No. 7, pp. 141-146. Hydrocarbon Process., July 1974.
18. Thompson, C. J., Dooley, J. E., Vogh, J. W., and Hirsch, D. E. "Analyzing Heavy Ends of Crude (Recluse)," Vol. 53, No. 8, pp. 93-98. Hydrocarbon Process., August 1974.

SEPARATION OF COAL LIQUIDS FROM MAJOR LIQUEFACTION PROCESSES TO MEANINGFUL FRACTIONS

I. Schwager and T. F. Yen
University of Southern California

Coal liquids have been separated by solvent fractionation into three crude fractions: pentane-soluble; pentane-insoluble and benzene-soluble (crude asphaltene); and benzene-insoluble. The pentane-soluble fraction has been further separated into a liquid propane-soluble fraction (gas oil), and a liquid propane-insoluble fraction (resin). The benzene-insoluble fraction has been further separated into a carbon disulfide-soluble fraction (carbene), and carbon disulfide-insoluble fraction (carboid). The great majority of these materials have been characterized by elemental and metal analyses, molecular weight, particle size, and color intensity determination, and NMR, IR, and x-ray diffraction analyses. The crude asphaltene fractions have been further separated into major fractions by solvent elution chromatography. The two major asphaltene fractions obtained from crude asphaltenes by elution from silica gel with benzene and diethyl ether differ markedly in nitrogen content indicating that species with basic nitrogens are being preferentially absorbed on silica gel. This suggests that silica gel chromatography may provide a mild, chlorine-free procedure for separating asphaltenes into acidic and basic components.

I. INTRODUCTION

The three direct general processes for converting coals to liquid fuels are: catalyzed hydrogenation, staged pyrolysis, and solvent refining (1,2). Each of these processes results in the production of a coal liquid which contains a variety of desirable and undesirable components. The desirable coal liquids are the oils-saturated and aromatic hydrocarbons plus nonpolar nonhydrocarbons, and the resins-polar nonhydrocarbons. The undesirable species are the asphaltenes and the carbenes-high molecular weight highly aromatic solids, and the carboids polymerized coke-like materials. The undesirable elements; metals, sulfur, nitrogen, and oxygen are generally present in higher concentration in the asphaltene and carboid fractions. Under hydrogenolysis conditions, the conversion of coal to oil has been suggested to proceed via the following sequence(3):

$$\text{Coal} \longrightarrow \text{Asphaltene} \longrightarrow \text{Oil}$$

Therefore, asphaltene generation and elimination are of great importance in the liquefaction process. A study of the chemical and physical properties of asphaltenes may lead to the discovery of ways to reduce or eliminate asphaltene build-up in coal liquids and to thereby increase the yields of desirable coal liquefaction products. In this work, coal liquids from representative liquefaction processes have been separated by solvent fractionation, and the fractions are being examined by various analytical and physical techniques. Particular attention is being directed toward asphaltene separation, purification and characterization.

II.   RESULTS AND DISCUSSION

A solvent fractionation scheme for separating coal liquid products into five fractions (oil, resin, asphaltene, carbene, and carboid) is shown in Fig. 1. Representative coal liquid samples produced via the three direct coal liquefaction processes were separated into the five fractions described above. The results are presented in Fig. 2. For the catalyzed hydrogenation product produced in the Synthoil process (4), the product composition is about 61% oil, 22% resin, 13% asphaltene, 0.6% carbene, and 3% carboid. The staged pyrolysis filtered product[1] from the FMC Corporation's COED process (5) has a product composition of about 26% oil, 48% resin, 15% asphaltene, 2% carbene and 10% carboid. The solvent refined coal (SRC) produced by Catalytic Inc. based on PAMCO's SRC process (6) affords about 4% oil, 15% resin, 45% asphaltene, 2% carbene, and 34% carboid. The results found in this work are in good agreement with those reported recently for solvent fractionation of a Synthoil catalytic hydrogenation product, and a non-catalytic SRC product (7).

---

1. The filtered product is the pyrolysis product produced prior to the final hydrotreating reaction. The FMC-COED Syncrude produced by hydrotreating the filtered product at 3100 psi, 775°F consists of about 99% oil, 0.8% resin, and 0.2% asphaltene.

SOLVENT FRACTIONATION SCHEME FOR COAL LIQUID PRODUCT

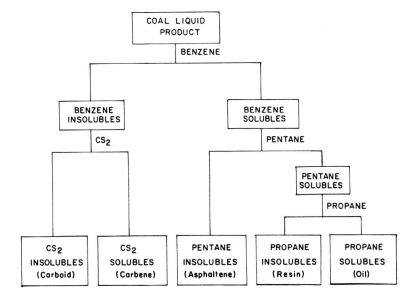

Figure 1   Solvent Fractionation Scheme for Coal Liquid Product

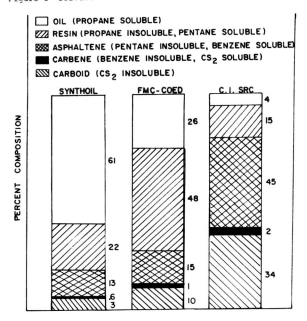

Figure 2   Solvent Fractionation Analysis of Coal Liquid Products

These workers found for a synthoil product, 80% pentane soluble material, 15% of pentane insoluble and benzene or toluene
soluble material, and 5% of benzene insoluble material. For
an SRC product they found 20%, 46%, and 34% respectively for
the above three fractions. They found their benzene insolu-
able fraction to be pyridine soluble, and named this fraction
'preasphaltenes' in the belief that it might be intermediate
between coal and classical asphaltenes. We feel, however,
that the benzene insoluble fractions may arise from reactive
coal depolymerization moieties which are not stabilized by
hydrogenation, but are repolymerized into materials more
difficult to degrade than the original coal substance. More
work will have to be done to elucidate the origin of this
fraction.

The analyses of the starting coal liquids and the various solvent fractions are given in Tables I, II, and III.
Semi-quantitative metal analyses are presented in Table IV.
It may be seen that heteroatoms and metals are generally
concentrated in the asphaltene and carboid fractions.

TABLE I
Synthoil Solvent Fractions Ultimate Analysis, %*

Fraction	C	H	N	S	O**	Ash	MW***
Coal liquid	87.26	8.44	0.94	0.10	3.26	0.69	
Oil	87.74	9.58	0.60	0.43	1.65	0.02	243
Resin	87.27	7.77	1.30	0.14	3.52	0.30	305
Asphaltene	87.27	6.51	1.63	0.66	3.93	0.48	738
Carbene	87.96	5.94	1.72	0.74	3.64	0.56	
Carboid	88.32	5.69	1.64	2.07	2.28	8.80	938****

*Moisture and ash free. **By difference. ***VPO in benzene
****VPO in DMF

Table V shows the color indices (8) (integrated absorp-
tion of a species between 750 nm and 400 nm) of the various
solvent fractions for different coal liquids. Since the
color index of an aromatic molecule is a function of the size
of the aromatic π-system, it appears reasonable to assume
that the increase in the color indices in going from oil, to
resin to asphaltene to carbene, reflects an increase in the
size of the respective π-systems.

Tables VI and VII show the carbon aromaticities, f_a (9)
and the hydrogen percentages by proton type (10) for asphaltenes produced in the different processes. Carbon aromaticities are found to increase in going from asphaltene to

carbene to carboid which is consistent with the color
indices results.

The solvent refined coals yield asphaltenes which contain a higher percentage of aromatic protons relative to
benzyl and saturated protons than do the other types of liquefied coals.

TABLE II
FMC-COED Solvent Fractions Ultimate Analysis, %*, MW

Fraction	C	H	N	S	O**	Ash	MW***
Coal liquid	83.04	7.68	1.09	1.11	7.08	0.28	
Oil	85.88	9.75	0.42	1.08	2.87	0.01	267
Resin	83.25	7.25	1.06	1.26	7.18	0.39	325
Asphaltene	82.14	6.47	1.70	2.58	7.11	0.79	458
Carbene	81.17	6.01	1.69	1.31	9.82	0.11	
Carboid	78.47	5.65	1.95	1.85	12.08	1.31	394****

*Moisture and ash free.   **By difference.   ***VPO in benzene.
****VPO in DMF.

TABLE III
Cat. Inc. SRC Solvent Fractions Ultimate Analysis, %*

Fraction	C	H	N	S	O**	Ash	MW***
Coal liquid	88.71	5.53	1.26	0.19	4.31	0.29	
Oil	90.99	6.94	0.41	0.57	1.09	0.25	264
Resin	89.88	6.64	0.84	0.00	2.64	0.05	370
Asphaltene	88.79	5.61	1.25	0.12	4.23	0.78	560
Carbene	89.77	5.03	1.11	0.26	3.83	0.40	
Carboid	87.08	4.70	0.96	0.34	6.92	0.71	1026****

*Moisture and ash free.   ** By difference.   ***VPO in benzene.
****VPO in DMF.

TABLE IV
Semiquantitative Metal Analysis*
==

Major Elements	Asphaltene	Benzene Insoluble
Synthoil		
Si	180	1800
Fe	130	420
Al	8	760
Ti	69	130
B	60	77
Ca	3	69
Mg	1	29
FMC-COED		
Si	210	100
Fe	20	270
Al	57	140
Ti	12	10
B	81	72
Ca	39	280
Mg	5	14
Cat. Inc. SRC		
Si	55	130
Fe	7	58
Al	6	420
Ti	1	71
B	5	18
Ca	26	96
MG	10	140

*Results in ppm.

Asphaltenes produced in the Synthoil process have been reported to consist of acidic and basic components (11). These components have been separated by treating the asphaltenes, dissolved in toluene, with dry HCl gas. The basic component precipitates as an HCl adduct in 57% yield, and the acidic and neutral components (43%) remain in solution. In this work we have further separated asphaltenes into two major fractions by solvent elution chromatography. The two major fractions obtained by elution from silica gel with benzene and diethyl ether have different properties although

they both may be freeze-dried to brown powders from benzene
solutions.

TABLE V
Color Indices, I* of Solvent Fractions

Fraction	Synthoil	FMC-COED	Cat. Inc. SRC
Oil	0.1	0.1	0.4
Resin	1.7	1.8	1.5
Asphaltene	13	4.1	17
Carbene	25	5.3	49
Carboid**	33	14	56

$*I = \int_{400 \text{ nm}}^{750 \text{ nm}} \text{Ad}\lambda$ in THF X 10^{-1} (relative).   **Measured in
Pyridine.

TABLE VI
Aromaticity, F_a*, for Asphaltenex, Carbenes and Carboids by
X-Ray Diffraction

Sample	f_a		
	Asphaltene	Carbene	Carboid
Synthoil	0.63**	0.63**	0.71**
FMC-COED	0.58**	0.81	0.84
Cat. Inc. SRC	0.67	0.72	0.83
PAMCO SRC	0.66	0.74	0.77

$*f_a = C_A/C_{total} = A_{002}/A_{002} + A_r$; C_A = number of aromatic C;
C_{total} = number of total C; A_{002} = area under peak for
aromatic carbons; A_γ = area under peak for saturated C.

**These values are less reliable due to overlap of back-
  ground because of use of aluminum sample holder. Other
  values determined by use of glass sample holder.

TABLE VII
NMR H-Percentages by Proton Type*

Asphaltene	$H_{aromatic}$	H_{benzyl}**	H_{sat}**
Synthoil	33	42	25
FMC-COED	37	43	20
Cat. Inc. SRC	50	34	16
PAMCO SRC	45	38	17

*Run on Varian T-60 NMR, Solvent 99.8% $DCCl_3$ + 1% TMS.
**Separation point between H_{benzyl} and H_{sat} chosen at $\tau = 8.27$

TABLE VIII
Silica Gel Chromatography of Asphaltenes

Asphaltene Sample	Total Wt % Recovered	Wt % With Solvent Benzene	Diethyl Ether
Synthoil	84	57	43
FMC-COED	88	38	62
Cat. Inc. SRC	89	66	34
PAMCO SRC	88	51	49

Table VIII gives the total weight % asphaltenes recovered and the % distribution of asphaltenes obtained using the eluents benzene and dithyl ether. Table IX compares the color indices of the solvent eluted fractions with those of the starting asphaltenes. The color indices for all the diethyl ether-eluted fractions are generally somewhat lower than either the starting asphaltenes or the benzene-eluted asphaltenes. This suggests a less extensive π-aromatic structure in these fractions. A comparison of the analyses and molecular weights of the various silica gel chromatography fractions is presented in Table X. The H/C ratios are highest in the diethyl ether-eluted fractions suggesting relatively more aliphatic carbons in this fraction.

TABLE IX
Color Indices, I*, of Silica Gel Fractions

Sample	Asphaltene	Eluent Benzene	Diethyl Ether
Synthoil	13	12	8
FMC-COED	4.1	5.1	3.5
Cat. Inc. SRC	17	13	11
PAMCO SRC	10	9	9

$*I = \int_{400\ nm}^{750\ nm} Ad\lambda$ in THF X 10^{-1} (relative).

TABLE X
Silica Gel Chromatography Fractions Ultimate Analysis, %*,
and MW

Sample							
Synthoil Asph.	87.27	6.51	1.64	0.66	3.92	0.48	738
Benzene	89.15	6.57	0.58	0.99	2.71	0.97	614
Diethyl Ether	85.17	6.86	1.60	0.73	5.64	1.19	560
FMC-COED Asph.	82.14	6.47	1.70	2.58	7.11	0.79	458
Benzene	83.83	6.30	1.06	1.93	6.88	1.38	445
Diethyl Ether	79.67	6.48	1.54	1.33	10.98	1.23	340
Cat Inc Asph.	88.80	5.61	1.25	0.12	4.22	0.78	560
Benzene	87.66	6.40	0.65	0.65	4.64	1.10	467
Diethyl Ether	84.73	6.33	1.67	0.79	6.48	1.78	490
PAMCO Asph.	86.33	6.15	1.47	1.13	4.92	1.50	432
Benzene	87.27	6.20	0.68	1.20	4.65	1.83	421
Diethyl Ether	83.00	6.52	1.99	1.01	7.48	2.03	465

*Moisture and ash free.   **By difference.   ***VPO in benzene.

The percentages of the heteroatoms nitrogen and oxygen are
also appreciably greater in the diethyl ether-eluted fractions. These results are consistent with the acidic nature

of silica gel, and the expectation that silica gel would preferentially adsorb basic molecules. This suggests that silica gel chromatography may provide a mild, chlorine-free procedure for separating asphaltenes into basic and non-basic components.

Actually, the words 'acid' and 'base' as reported by Steinberg, et al, (11,7) do not adequately classify or clarify the coal-derived asphaltene. We would suggest acceptor (π-deficient) and donor (π-abundant), since their association and the nature of charge transfer is well known in asphaltene (12). The asphaltene thus formed is a complex and not as a salt in the sense of ionizable species of acid or base. We feel the benzene-eluted fraction could be neutral, the ethyl ether-eluted fraction polar, and by properly selecting adequate solvent, another more polar fraction can also be obtained. All these phenomena can be explained on the basis of charge-transfer. Actually, the existence of different degrees of association of this charge-transfer makes the task of separation of asphaltene difficult.

III. SUMMARY

A preliminary examination of coal liquefaction products from four different coal liquefaction processes has been carried out. Each coal liquid has been separated into five different fractions by solvent fractionation. Total recoveries ranging from 93 to 97% by weight have been obtained. These solvent fractions are respectively: oil, resin, asphaltene, carbene, and carboid. We have further separated the asphaltene fraction by utilization of solvent elution chromatography with silica gel, into two fractions of different polarities. Unlike the asphaltene separation method described by Sternberg, et al, (11), which introduces chlorine chemically into at least one of the separated components, the present method does not chemically alter the asphaltenes.

IV. ACKNOWLEDGEMENTS

The work described here was sponsored by ERDA E (49-18) 2031, to whom the authors wish to express their appreciation.

V. LITERATURE CITED

1.   Burke, D. P., Chem. Week, 115, 38 (1974).
2.   Cochran, N. P., Sci. Amer., 234, 24 (1976).

3. Weller, S., Pelipetz, M.G., and Friedman, S., Ind.Eng. Chem., 43, 1572, 1575 (1951).

4. Sternberg, H. W., Raymond, R., and Akhtar, S., Preprints, Div. Petrol. Chem., ACS, 20, 3, p. 711 (1975).

5. FMC Corporation, Char Oil Energy Development, Research and Development Report No. 56, Interim Report No. 1, Office of Coal Reasarch, U. S. Dept. of the Interior (1970)

6. Pittsburgh and Midway Coal Mining Co., Economics of a Process to Produce Ashless, Low Sulfur Fuel from Coal, Research and Development Report No. 53, Interim Report No. 1, Office of Coal Research, U. S. Dept. of the Interior (1970).

7. Sternberg, H. W., Preprints, Div. Petrol. Chem., ACS, 21, 198 (1975).

8. Yen, T. F., Erdman, J. G., and Saracero, A. J., Anal. Chem., 34, 694 (1962).

9. Yen, T. F., Erdman, J. G., and Pollack, S. S., Anal. Chem., 33, 1587 (1961)

10. Brown, J. K. and Ladner, W. R., Fuel, 39, 87 (1960).

11. Sternberg, H. W., Raymond, R., and Schweighardt, F. K., Sci., 188, 49, (1975).

12. Yen, T. F., Fuel, 52, 93 (1973).

HIGH PRESSURE LIQUID CHROMATOGRAPHIC
STUDIES OF COAL LIQUEFACTION KINETICS

John W. Prather, Arthur R. Tarrer, James A. Guin,
Donald R, Johnson, and W. C. Neely
Auburn University

High pressure liquid chromatography (HPLC) provides
a relatively simple analytical method for analysis of
the complex organic mistures found in coal liquefaction
processes. This technique offers the advantage that
preparatory scale work is accomplished with relative
ease allowing for positive identification of the various
components by other methods, e.g., infrared and ultraviolet spectroscopy. The feasibility of using HPLC to
characterize solvents used in the Solvent Refined Coal
(SRC) process is reported. The effects of catalytic
agents - namely, a commercial Co-Mo-Al catalyst; a coal
mineral, iron pyrite; coal ash; and actual mineral residue from an SRC process on twelve constituents of a coal
derived solvent, creosote oil - is monitored using HPLC.

I. INTRODUCTION

A major obstacle to the advancement of coal conversion
technology has been the lack of effective means for analyzing
the highly complex mixtures occurring in coal conversion processes. High pressure liquid chromatography (HPLC) is a
recently developed analytical tool that typically provides
rapid, reproducible analysis of complex system (1,2). Using
HPLC, preparatory experiments can be performed to separate
50-100 mg samples of relatively pure compounds from a complex
mixture - allowing subsequent positive identification of the
compounds by other means such as infrared or ultraviolet
spectroscopy, and melting point. This asset of HPLC (i.e.,
preparatory scale chromatography) is one of the main advantages that it offers over comparable analytical methods.
Because HPLC does offer positive identification of individual
species, it is a very effective tool for characterizing and

quantifying process streams in coal liquefaction processes, and provides a means for developing a more fundamental understanding of such processes.

The present work demonstrates the feasibility of employing HPLC to characterize creosote oil, a coal-derived liquid used as a start-up solvent in coal liquefaction processes such as the Solvent Refined Coal (SRC) process. Characterization of the creosote oil is done by HPLC during hydrogenation/hydrodesulfurization of the oil; and is then used to follow the catalytic effects of a commercial Co-Mo-Al catalyst; a coal mineral, iron pyrite; coal ash; and actual SRC mineral residue from the Wilsonville pilot plant. Each of these agents has a significant effect on the hydrogenation and, except for pyrite, on the hydrodesulfurization of creosote oil under conditions similar to those in the SRC process. Evidence that coal minerals have a catalytic effect on hydrogenation of coal has been reported (3,4). However, specifically which compounds in recycle or process solvent are most affected by hydrogenation in the presence of coal minerals has not been previously studied. Also the effect of coal mineral catalysis on rate of removal of various heteroatom compounds - particularly sulfur bearing compounds such as dibenzothiophene - has not been studied. HPLC provides a tool for monitoring changes in major constituents of coal-derived liquids during hydrogenation/hydrodesulfurization, and can be used to study the selectivity of catalytic agents for accelerating reactions involving specific species of coal-derived liquids - as is demonstrated here.

II. EXPERIMENTAL

A. Apparatus

A Model ALC/GPC-201 high speed liquid chromatograph (Waters Associates, Milford, Mass.) was used throughout this study. The following accessory hardware was used: a Model 6000 solvent delivery system, a Model 660 solvent programmer, and a Model U6K injection system. In addition, a Model GM77 UV/VIS detector (Schoeffel Instrument Corporation, Westwood, N.J.) was used; this detector was chosen because it has a continuously variable UV source. After thorough exploratory studies, a wavelength setting for the detector of 232 nm was found to provide the best overall sensitivity and stability for detection; this setting was used throughout the study.

B. Reagents

The acetonitrile used in this study was of spectroquality.
The creosote oil was obtained from Southern Services, Inc.,
at the SRC pilot plant located at Wilsonville, Alabama.
Southern Services, Inc., obtained the oil, creosote oil 24-
CB, from the Allied Chemical Company. The oil has a boiling
point range of 175° to 400^oC and a specific gravity of 1.096
at 20^oC. Hydrogen was obtained from Linde Hydrogen in 6000
psi grade, with a purity of 99.995%.

A commercial Co-Mo-Al catalyst (Comox-451) was obtained
from W. R. Grace and Company, Davidson Chemical Division,
Baltimore, Maryland. This catalyst is commercially produced
by Laporte Industries of England. Our analysis of the catalyst showed that it consists of 3.7% CoO and 12.8% MoO_3, and
the catalyst was specified by the manufacturer to have a
surface area of 300 m^2/g and a total pore volume of 0.66 ml/g.
The pyrite used in these experiments was obtained from Matheson Coleman and Bell Chemical Company, Norwood, Ohio. Our
analysis of the pyrite showed that it was 90-95% pure, the
difference being primarily silica. Coal ash was obtained by
burning a mixture of Kentucky No. 9/14 coal mixture (7.2% ash)
in a muffle furnace at 1000^oC. Analysis of the ash gave an
iron content of 13.7%. SRC solids were obtained from the
filter cake from the Wilsonville, Alabama SRC Pilot Plant.
Analysis of the material gave an ash content of 55.2% and a
sulfur content of 13.6%. The reported analysis of this material showed that it was 30% filter aid (diatomaceous earth)
(5). All materials were ground and screened; and only -325
mesh (45 micron) size material was used.

C. High Pressure Liquid Chromatography

Two 4mm (ID) X 30 cm microbondapak/C_{18} columns (Waters
Associates, Milford, Mass.) were used in series for separation
of the creosote oil components. The mobile-phase was a 45:55
volume-to-volume acetonitrile-water mixture. The flow rate
was nonlinearly (curve 8 on the 660 solvent programmer) programmed to increase from 0.6 to 0.8 ml/min in 1.5 hours in
such a way that the major portion of the increase takes place
in the last 45 minutes of the program. This allows for the
most efficient use of the time necessary for elution by minimizing dead-time and axial diffusion - while keeping resolution at a maximum during the initial portion of the separation.
System pressure was kept below a maximum of 2200 psi.

The sample was dissolved in pure acetonitrile (4mg/ml)
prior to injection, and normally about 10 μl of the resulting

solution was injected for analysis. Retention times ranged
from as short as a few minutes for the more polar compounds
to as much as several hours for the more nonpolar compounds,
which are typically the higher molecular weight constituents.

D. Procedure

Creosote oil was treated for two hours at 425°C under an
initial pressure of 3000 psig of hydrogen, in the presence of
13% by weight of either Co-Mo-Al catalyst, iron pyrite, coal
ash, or SRC solids. Treatment was also made without any
catalytic agents present. The reaction mixtures were stirred
continuously at 2000 rpm in a batch autoclave (Autoclave Engineers, Erie, PA.). After two hours of reaction, samples of
the partially hydrogenated and hydrodesulfurized oil were
taken. Then, aliquot amounts of these samples were dissolved
in acetonitrile (4mg/ml), filtered through 0.5μ filters (Milli-
pore Intertech, Inc., Bradford, MA.) to remove any solids; and
after filtration, 10μl portions of the filtrate were analyzed
by HPLC. To eliminate unavoidable variations due to differences in the amount of sample injected, benzofuran was
used as an internal standard.

E. Computer Deconvolution

As is seen in Figure 1, the chromatograms obtained in
this work gave peaks that seldom had base-line resolution - a
common problem in chromatography - and, to be accurate, it was
necessary to deconvolute the chromatograms. To this end, a
specialized computer program was developed (6) based on the
method of Marquardt (7).

III. RESULTS AND DISCUSSION

Typical chromatograms of the product from the various
treatments of the creosote oil are shown in Figure 1. Good
separation of the multicomponent system was obtained by
varying the acetonitrile/water ratio in the eluting solvent:
the best resolution resulted with a 45:55 volume-to-volume
acetonitrile-water mixture. Preparatory scale columns were
used to obtain sufficient amounts of twelve major constituents
to permit their positive identification by infrared spectroscopy. These major constituents, which compose 50.3% by weight
of the original creosote oil, were of interest in this preliminary work (Table 1).
The components fall into basically two categories: polynuclear aromatics and heteroatom compounds. Compounds of the

FIGURE 1. Chromatograms of creosote oil which show the effect of the various treatments. Postively identified components in order of elution are: a) internal standard (Benzofuran), b) 1 and 2-naphthonitrile, c) carbazole, d) naphthalene, e) 2-methylcarbazole, f) 1-methylnaphthalene, g) 2-methylnaphthalene, h) dibenzofuran, i) biphenyl, j) acenaphthene, k) fluorene, l) dibenzothiophene, m) phenanthrene, n) anthracene.

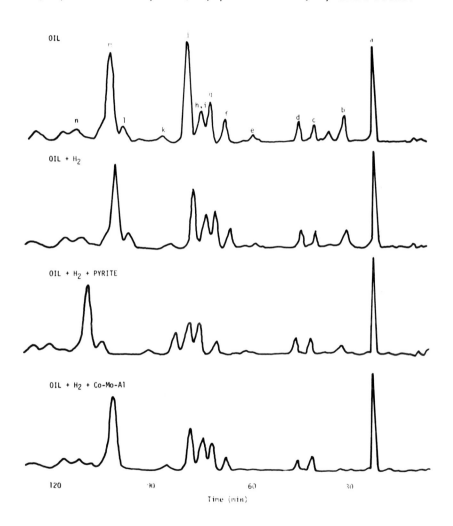

TABLE 1
High Pressure Liquid Chromatographic Partial Analysis of
Recycle Solvent

Compound	Weight %
1 and 2-naphthonitrile	0.03
carbazole	0.34
naphthalene	7.27
2-methylcarbazole	0.11
1-methylnaphthalene	2.64
2-methylnaphthalene	7.54
fluorene	3.10
acenaphthene	3.10
dibenzothiophene	0.59
phenanthrene	8.42
anathracene	1.37

TABLE 2
Comparison of Creosote Oil Analysis

Compound	A	B	C	D
		Wt.%		
1 and 2-naphthonitrile	0.32	--	0.61	--
carbazole	2.2	--	0.42	--
naphthalene	5.1	9.08	8.92	10.0
2-methylcarbazole	1.7	--	0.11	--
1-methylnaphthalene	0.38	3.55	5.23	3.0
2-methylnaphthalene	1.3	10.2	8.00	8.0
acenaphthene	6.0	9.73	6.28	5.0
fluorene	10.3	5.54	5.22	5.0
dibenzothiophene	0.52	0.94	1.01	--
phenanthrene	18.6	17.92[a]	12.45	17.0[a]
anthracene	4.3		1.86	

A - Allied Chemical Company by gas chromatography
B - Southern Services by gas chromatography (ASTM D2887)
C - Auburn University by HPLC
D - Auburn University by gas chromatography

a - includes both phenanthrene and anthracene

first type - namely, naphthalene, acenaphthene, phenanthrene, and anthracene - are angular polycyclic aromatics. These compounds, and compounds like them, are considered to play an important role in hydrogen transfer reactions occurring during coal liquefaction, e.g. as in the SRC process (8,9,10). Compounds of the second type - namely, dibenzothiophene, carbazole and naphthonitrile - are of interest because these constituents, during combustion of coal, form pollutants, SO_2 and NO_x. Determination of the degree to which these constituents are removed with processing is therefore most important for process analysis.

The results from this work and those from previous analysis using gas chromatography are compared in Table 2. In evaluating this comparison, it should be noted that the creosote oil is somewhat unstable: after standing for long periods of time (several months) some constituents do indeed precipitate from solution. For this reason some discrepancies exist between our gas chromatographic analysis, that performed by Allied Chemical Company, and that performed by Southern Services.

Table 1 lists the analysis of the creosote oil after the various treatments. Obviously, hydrogenation and hydrodesulfurization of the creosote oil at $425^{o}C$ in the presence of an initial hydrogen partial pressure of 3000 psig causes a significant decrease (20%) in percent by weight of the analyzed components. Currently, preparatory scale work is in progress to determine what compounds are produced by this treatment. In the presence of the various catalytic agents, even more reduction (34%, when pyrite is present, to 49%, when actual SRC mineral residue is present, as compared to 20%, when no catalyst is present) in the weight percent of the analyzed components results during hydrogenation of the oil. Apparently then, these agents (Co-Mo-Al, pyrite, coal ash, and SRC mineral residue) do indeed catalyze hydrogenation of the oil, and the increase in hydrogen consumption observed when they are present is, in fact, due to greater hydrogenation of the oil, rather than reduction of the agent, itself, with hydrogen to produce a reduced form of the agent, water, hydrogen sulfide, etc.

As shown in Table 1, HPLC offers sufficient specificity to detect differences in the final concentration of major constituents of the oil when hydrogenated in the presence of the different catalytic agents. For example: assuming that the disappearance of the major constituents is due to hydrogenation, then Co-Mo-Al, coal ash, and SRC mineral residue show a decisive preference for accelerating hydrogenation of naphthalene, 1-methylnaphthalene, and 2-methylnaphthalene; whereas pyrite favors hydrogenation of acenaphthene and, to a lesser extent, anthracene. Also, in the presence of commercial Co-Mo-Al catalyst, the concentration of dibenzothiophene, a major organic sulfur component, is reduced to a much greater extent than when the other catalysts were present. In fact, the concentration of dibenzothiophene was reduced essentially to zero when Co-Mo-Al was present. Since Co-Mo-Al is an excellent catalyst for hydrodesulfurization reactions, these results are not surprising. Most interestingly, however, the trend in dibenzothiophene removal is exactly the same as that found by analysis of total sulfur (4): namely, Co-Mo-Al >> coal ash > SRC mineral residue > pyrite > H_2 only. Carbazole, on the other hand, has essentially the same concentration as that in the original oil, despite the catalyst used; whereas the naphthonitriles are completely removed when either Co-Mo-Al, coal ash, or SRC mineral residue is present.

IV. CONCLUSIONS

The HPLC procedure described here permits analysis of major constituents of creosote oil in about two hours. No extensive sample preparation is required; and the method is quantitative - the results comparing well with those obtained using gas chromatography. Using preparatory chromatography, samples of essentially pure compounds can be separated from creosote oil and positively identified by subsequent analysis. As a result, with HPLC neither tentative identification as, for example, by "spiking" - i.e. addition of known compounds to the mixture so as to determine the peaks in the chromatogram of the mixture caused by different constituents - nor an expensive, sophisticated gas-chromatograph/mass-spectrometer system are required. Hence HPLC offers a simple and powerful technique for analysis of complex mixtures like creosote oil.

V. ACKNOWLEDGMENTS

The research reported here is supported by the RANN division of the National Science Foundation under Grant No. 38701, by Auburn University Engineering Experiment Station, and by the Alabama Mining Institute. The authors are especially indebted to Southern Services, Inc., for supplying various materials and to W. R. Grace Co. for supplying Co-Mo-Al catalyst.

VI. LITERATURE CITED

1. C. Zweig and J. Sherma, *Anal. Chem.*, 46, 73R (1974).
2. W. A. Dark, *Amer. Lab.*, August, 50 (1975).

3. C. H. Wright and D. E. Severson, ACS, Div. Fuel Chem. Preprints, 16 (2), 68 (1972).
4. A. R. Tarrer, J. A. Guin, J. W. Prather, W. S. Pitts, and J. P. Henley, ACS, Div. Fuel Chem. Preprints, This volume (1976).
5. Status Report of Wilsonville Solvent Refined Coal Pilot Plant, prepared by Southern Services, Inc., Birmingham, Alabama, for the Electric Power Research Institute, EPRI interim report 1234, May, 1975.
6. W. S. Pitts, Department of Chemical Engineering, Auburn University, Auburn, Alabama, Personal Communication.
7. D. M. Marquardt, J. Soc. Indust. Appl. Math, 11, 431 (1963).
8. H. G. J. Potgieter, Fuel, 52, 134 (1973).
9. L. A. Heredy and P. Fugassi, "Coal Science", A.C.S. Advances in Chemistry Series, 55, 448 (1966).
10. E. E. Donath, "Chemistry of Coal Utilization," H. H. Lowry, Ed., John Wiley: New York, 1963, Chap. 22, p. 1063.

A PETROGRAPHIC CLASSIFICATION OF SOLID RESIDUES DERIVED FROM THE HYDROGENATION OF BITUMINOUS COALS

Gareth D. Mitchell, Alan Davis, and William Spackman
The Pennsylvania State University, Coal Research Section

A classification has been established for solvent-extracted filter-cake solid residues of high volatile bituminous coals from several bench-scale and pilot plant solvent-refined coal operations. An attempt to relate the organic residue constituents with the precursor coal macerals, and to distinguish between potentially inert and reactive materials has provided insight into the physical and chemical degradation of coal during hydrogenation. The proposed classification is based on morphology, relative reflectance and anisotropy, and size, as determined by examination of polished surfaces in reflected white light.

The experimental approach involved the batch hydrogenation of selected coals through a range of temperature (300-450°C, at 25°C increments). The benzene-insoluble residues from each series were studied optically and provided a record of degradation for each coal maceral. The resulting phases were compared to residues derived from pilot plant and bench-scale systems.

The microscopic examination of residues could provide a routine method for determining the instantaneous efficiency of continuous-flow reactors, should aid in selection of feedstocks for possible future conversion, and to some extent will provide a critical analysis of the reaction conditions.

I. INTRODUCTION

Previous optical studies of coal dissolution have shown that the grain size of the organic constituents decreases with increasing reaction temperature (1)[1]. These earlier investigations were concerned mainly with the chemical aspects of coal hydrogenation as a function of hydrogen transfer and thermal degradation; however no attempt was made to relate these parameters to the changing physical character of coal macerals. The "char" components produced during hydrogenation are known to be optically dissimilar to their original coal constituents. Therefore, an understanding of the physical mechanisms involved in the transformation of coal macerals into these "char" or residue constituents is necessary in order to determine some of the chemical aspects of coal hydrogenation.

A characterization of the morphology of solid residues and consequently their classification, is the principal basis for studying the physical mechanisms of coal maceral degradation. However, the precursors of residue components can not be determined simply by comparing the petrography of a feed coal to that of its solid residue. These relationships may be better understood by observing the changes in the nature of residues produced during progressive batch hydrogenation at successively higher temperatures. These studies have served to relate the original feed coal macerals to residue components derived from continuous-flow reaction.

Optical studies reveal that residue constituents are formed as a result of incomplete hydrogenation, repolymerization or carbonization. Those coal macerals which pass through the process unreacted or exhibiting only slight alteration are readily identified. Other constituents are related to their precursor macerals by studying low temperature batch hydrogenation residues and are described by their morphology, relative reflectance, anisotropy, and grain size.

II. SAMPLE ORIGIN

Residues from two coal hydrogenation systems were used in this study. An experimental batch hydrogenation system

---

[1]Given, P.H., W. Spackman, A. Davis, P.L. Walker, Jr., and H.L. Lovell. 1975. The relation of coal characteristics to coal liquefaction behavior. Unpub. Report College of Earth and Mineral Sciences, The Pennsylvania State University to National Science Foundation, no. 1, March, 43 p.

utilized 20 g of minus 20 mesh coal with four parts by weight
of tetralin in a stainless steel reaction vessel (nitrogen
atmosphere). These preparations were reacted in a fluidized
sand bath for three hours at seven constant temperatures in
the range of 300-450°C at 25°C increments[1]. The reaction
vessel contents were extracted with benzene and the resulting
insoluble fractions were dried, embedded in epoxy resin, and
polished.

The second system employed was a proprietary continuous-
flow bench-scale reactor[2]. This process used a feed rate of
400 g/hr of minus 200 mesh coal slurried in a 1 to 3 ratio
with solvent-recycle or anthracene start-up oil. The slurry
was pumped under a pressure of 1500 psig into a reaction
column with hydrogen. Reported reaction temperatures of 441
and 427°C were given for a residence time approximating 30
minutes. Filtration was conducted at regular intervals to
determine the product balance for each run. Filter cake materials were extracted with pyridine and the resulting residues were dried, embedded in epoxy resin, and polished.

A pilot plant[3] version of the second system also provided materials that were used in this study. Solids which
were deposited in the reactor were subjected to microscopic
observation to determine their contribution to the insoluble
residues. These materials were washed with benzene and prepared for microscope examination as before.

---

[1]Given, P.H., W. Spackman, A. Davis, P.L. Walker, Jr., and
H.L. Lovell. 1974. The relation of coal characteristics to
coal liquefaction behavior. Unpub. Report College of Earth
and Mineral Sciences, The Pennsylvania State University to
National Science Foundation, no. 3, September, 74 p.

[2]Bench-scale "Solvent Refined Coal" unit, sponsored by the
Electric Power Research Institute and Hydrocarbon Research,
Inc. and operated by Hydrocarbon Research, Inc., Trenton,
New Jersey Laboratory.

[3]"Solvent Refined Coal" pilot plant at Wilsonville, Alabama,
sponsored by the Electric Power Research Institute and
Southern Service, Inc. and operated by Catalytic, Inc.

III. RESULTS AND DISCUSSION

A. Petrography of Feed Coals

The petrography of feed coals used for conversion in these hydrogenation systems are summarized in Table 1. The more reactive coal constituents, vitrinite and exinite (see Fisher *et al.*(2) and Davis *et al.*(3)) appear in varying concentrations. The reactive maceral content of the Indiana #6, West Kentucky #9/14, and Illinois #6 feed coals are relatively high while those for the Indiana #1 Block and the Australian Callide coals are substantially lower. These coals are high volatile bituminous in rank with the exception of the subbituminous A Callide seam.

B. Vitrinite Contributions to Residues

Table 2 presents the proposed residue classification and mechanisms which link the residue components to the macerals from which they were derived. The most diverse contribution to hydrogenation solid residues is obtained from vitrinite. The fact that this maceral exhibits a capacity to swell or contract within the temperature range of liquefaction is well documented. However, the appearance of unreacted vitrinite in these residues may be inconsistent with the thermoplastic properties of this maceral. Fig. 1 shows a particle of vitrinite containing a coalified spore exine (sporinite) filtered from the bench-scale process after reaction of the West Kentucky #9/14 feed coal at 441°C. Typically, the vitrinite in these residues shows some thermal alteration (parting) along bedding planes. However, the particle illustrated here attained only a slight granularity along one of its edges. The apparent lack of thermal degradation and incomplete hydrogenation of vitrinite suggest that the reaction temperature and/or residence time were not sufficient for reaction to occur.

Fig. 2 shows two additional vitrinite-residue components formed during the batch hydrogenation (at 350°C) of the Indiana #1 Block feed coal. The appearance of particles with

TABLE 1

PETROGRAPHIC ANALYSES OF FEED COALS

Sample No.	Method of Hydrogenation	Seam	State	Macerals, Vol. %							
				Vit.	Fus.	Sf.	Mac.	Mic.	Ex.	Res.	Cut.
PSOC-106	Exp.[1]	Indiana #1 Block	Ind.	25	10	12	21	6	23	2	1
PSOC-280	Exp.	Indiana #6 Seam	Ind.	92	1	1	0	3	2	1	0
PSOC-303	Exp.	Callide Seam	Queensland, Australia	15	4	73	1	2	2	0	3
POC-288	Continuous-Flow[2]	W.Ky.#14 Seam	Ky.	74	3	5	3	13	2	0	0
POC-289	Continuous-Flow	Ill. #6 Seam	Ill.	91	5	2	0	1	1	0	0

[1] Experimental batch hydrogenation
[2] Continuous-flow reaction

Vit. - vitrinite   Mac. - macrinite   Res. - resinite
Fus. - fusinite   Mic. - micrinite   Cut. - cutinite
Sf. - semifusinite   Ex. - exinite

259

TABLE 2

BEHAVIOR OF COAL ORGANIC CONSTITUENTS DURING LIQUEFACTION

Maceral Precursor	Mechanisms of Reaction		Organic Residue Components
Vitrinite	Slightly altered vitrinite (contracted and/or swollen) Granular residue Vitroplast (high molecular weight) → cenosphere Vitroplast (intermediate molecular weight) → vitroplast Hydrogenated product (repolymerized) → liquid-crystal ⇄ mesophase → semi-coke	} semi-coke	Unreacted vitrinite Granular residue Cenosphere Vitroplast Semi-coke (anisotropic)
Exinite	Fractional contribution to submicron granular material		Granular residue
Semifusinite	Possible fractional contribution to liquid product, mechanism similar to vitroplast		Semifusinite, semi-coke, or vitroplast
Fusinite	No observable mechanism		Fusinite
Macrinite	Possible fractional contribution to liquid product, mechanism unknown		Unknown
Micrinite	Mechanism unknown		Unknown

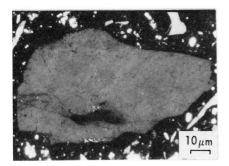

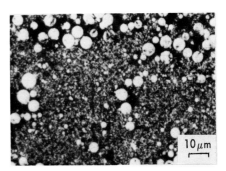

Fig. 1. Unreacted vitrinite containing sporinite from continuous-flow reaction of the West Kentucky #9/14 (441°C).

Fig. 2. Vitroplast and granular residue produced in batch hydrogenation of the Indiana #1 Block at 325°C.

a spherical morphology which is here called vitroplast[1] and a submicron material which has been referred to as granular residue are commonly seen in hydrogenation residues (Table 2). The term "vitroplast" describes a plastic or once-plastic degradation product of vitrinite which, unlike its maceral precursor, is optically isotropic. In one form it is characterized by flow structures as well as the spherical morphology seen in Figs. 2, 4, and 5. Vitroplast also appears as broad (>100µm) areas with inclusions of other residue components, and as angular fragments. Its identification in continuous-flow residues is often difficult due to a gradual transition into an anisotropic semi-coke.

The dissolution of vitrinite during the hydrogenation of the Indiana #6 feed coal at 325°C is seen in Fig. 3. The transition from slightly thermally-altered vitrinite into the vitroplast with flow structure and spherical morphology can be clearly observed in this photomicrograph. The lower reflectance of these vitroplast structures in comparison to the remnant vitrinite (Fig. 3) is an indication of the disorientation of the vitrinite lamellae. The transition of vitrinite into vitroplast may also be observed in residues derived from continuous-flow reaction, an example of which is illustrated in Fig. 4 (West Kentucky #9/14 residue at 441°C).

---

[1]Vitro - derived from vitrinite, refers to the original material and plast - derived from the Greek *plastos*, formed, molded.

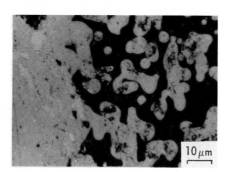

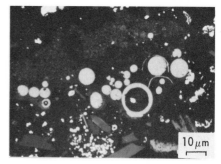

Fig. 3. Vitrinite
plasticity and the formation
of vitroplast at 325°C, after
batch hydrogenation of the
Indiana #6 feed coal.

Fig. 4. Vitroplast and
a simple cenosphere in the
West Kentucky #9/14 residue
(441°C, continuous-flow
reactor).

The granular residue seen in Fig. 2 is submicron in size
(0.3-1.0μm), apparently spherical, and appears blue and
red in polarized reflected light. During progressive hydrogenation, a similar granular material increases in concentration within the structure of the lower-reflecting vitrinite
species as each successively higher reaction temperature is
attained. This situation resembles the genesis of micrinite
from weakly reflecting vitrinite during metamorphic coalification as discussed by Teichmüller (4). Qualitative electron
probe analysis of the submicron constituents from a continuous-flow residue (Illinois #6 feed coal) indicates that only
a small portion of the particles tested were organic. The
remaining particles, predominately aluminosilicates, are
consequently assumed to represent fragments of finely dispersed clay. More work is required before the quantitative
distribution of organic and inorganic constituents of this
submicron fraction can be determined.
    The further alteration of vitrinite and the spherical
vitroplast derived from it is seen in the development of
"cenospheres". This term was first applied to structures
formed as a result of rapid heating of pulverized coal by
Newall and Sinnatt (5) and later by Street et al.(6). Cenosphere is a morphological term defined as a reticulated
hollow sphere composed of ribs or frames and windows. In
hydrogenation residues simple cenospheres which lack reticulate texture and the thin membranes or windows are often
observed (3,7). These less complex structures are observed

in the continuous-flow residue of the West Kentucky #9/14
feed coal (Fig. 4). Simple cenospheres are seen in close
association with the spherical vitroplast, and have similar
morphologies. The relationship between these residue species
may be observed in the development of more complex cenospheric
structures. Fig. 5 suggests the development of gas bubbles
in the interior of a vitroplast sphere from the continuous-
flow residue of the West Kentucky #9/14 coal (427°C). While
in the plastic state, gases formed by thermal cracking exert
sufficient pressure to cause expansion and eventually lead to
the formation of a highly reticulate hollow sphere as seen in
Fig. 6.

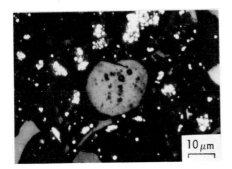
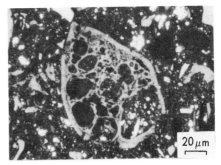

Fig. 5. Vitroplast ex-
hibiting gas bubble generation as a prelude to cenosphere formation. Continuous-flow reaction of West
Kentucky #9/14 (427°C).

Fig. 6. Highly reticu-
late cenosphere with carbonized outer wall. Continuous-
flow reaction of the West
Kentucky #9/14 at 441°C.

Microscopic observation reveals that as vitrinite is
subjected to increasing reaction conditions it may swell and
become plastic, forming both a fine-grained residue fraction,
inert to further hydrogenation, and a plastic, low-viscosity
phase (vitroplast) which is immiscible with the hydrogen-
donating solvent. The granular residue constitutes a major
portion (>50%) of these residues and may be considered as a
semiquantitative category composed of both inorganic and
organic constituents. The vitroplast is more variable in
concentration, much like the "unreacted vitrinite" category.
Quantitative data will be required to distinguish whether
these varying concentrations are due to characteristics of
the feed coal or to the reaction conditions.
As vitrinite physically disintegrates and loses its
molecular ordering in the initial stages of vitroplast

formation, there is a decrease in its reflectance. The
spherical vitroplast seen in continuous-flow residues
(Fig. 4), however, generally has a reflectance well above
that of vitrinite in the original feed coal. Presumably,
the disorientation of the vitrinite lamellae resulting from
plasticity, the subsequent immiscibility of the plasticized
vitrinite with the surrounding vehicle oil, and the loss of
internal pore structure makes hydrogenation increasingly
difficult. The recombination of thermally ruptured bonds (7)
formed during plasticity may result in the formation of a
higher reflecting, carbonized spherical vitroplast. A further increase in temperature may result in devolatilization
of the spherical vitroplast, causing internal thermal cracking and expansion to form cenospheres.

During hydrogenation repolymerization of the vitrinite-
derived liquids may occur as a result of hydrogen starvation,
erratic temperature fluctuation, pressure drops, or a critical accumulation of solids. This observation is confirmed by
the presence of semi-coke formed specifically by the mesophase mechanism. Mesophase is a transient intermediate stage
between a unique fluid system (liquid crystals) of high
molecular weight and a solid anisotropic carbon such as semi-
coke (8). Marsh *et al.* (9) have shown that liquid crystal
development can occur in the carbonization of some coals and
coal tar pitches, suggesting that during liquefaction, coal-
derived substances can repolymerize to form mesophase. The
initial appearance of mesophase can be an anisotropic sphere
formed by the lamellar stacking of the liquid-crystal poly-
mers parallel to the sphere's equatorial plane. A slight
increase in temperature can result in this transient phase
being converted irreversibly into semi-coke. A second mode
of occurrence of mesophase which is of greater significance
to liquefaction processes is the adsorption of nematic liquid
crystals on inert surfaces.

A potentially deleterious effect of mesophase formation
during hydrogenation is seen in Fig. 7. During a continuous-
flow pilot plant reaction of the Illinois #6 feed coal,
mesophase is shown to have formed in concentric layers on the
surface of an inert particle. Whether polymerization oc-
cured at the surface or within the surrounding fluid can not
be determined. However, as a result of lamellar stacking and
the development of cross-linking, an anisotropic nucleated
carbon is formed which is inert to further hydrogenation.
This mechanism of nucleation may serve to increase the specific gravity of these particles and promote their retention in
the reaction vessel.

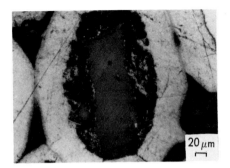

Fig. 7. Mesophase nucleation on calcite causing semi-coke formation. Pilot plant continuous-flow reaction of Illinois #6 feed coal at 446°C.

Fig. 8. Sporinite observed in the Indiana #1 Block feed coal, showing evidence of micrinite formation during normal coalification.

C. Liptinite Contributions to Residues

The liptinite group of macerals includes sporinite, resinite, cutinite, and alginite. Of these, sporinite often appears in low concentrations in high volatile bituminous coals. Owing to the very minor concentrations of the other liptinite components, it is not necessary to consider them here.

Teichmüller (4) suggests that micrinite forms from sporinite and other liptinite macerals during the metamorphosis of coal. She contends that the genesis of spore-derived micrinite occurs only during the high volatile bituminous stage of coalification. This suggests the possibility that sporinite might break down into a granular residue constituent as well as liquid by-products during hydrogenation. Fig. 8 exhibits a megaspore exine from the Indiana #1 Block coal which, as indicated by its porous and granular surface, may be forming this granular material during the course of coalification. Fig. 9 is a photomicrograph of the remains of a microspore after batch hydrogenation of the same feed coal at 375°C. A high reflecting granular material is observed along its swollen and partially reacted central cavity. One implication is that during hydrogenation of high volatile bituminous coals, a submicron granular material may be generated from sporinite to become a possible addition to other inert constituents.

D. Inertinite Residue Contributions

The group of coal macerals known as inertinite includes fusinite, semifusinite, macrinite, and micrinite. Their inertness relative to other macerals was determined through their behavior in carbonization; however, the term "inert" does not fully describe their properties. Some of these macerals are considered to be semi-inert even in coke making (10) and the same may be true for liquefaction.

Fig. 9.   Remnant micro-spore exhibiting granular residue formation in the 375°C batch hydrogenation residue of the Indiana #1 Block.

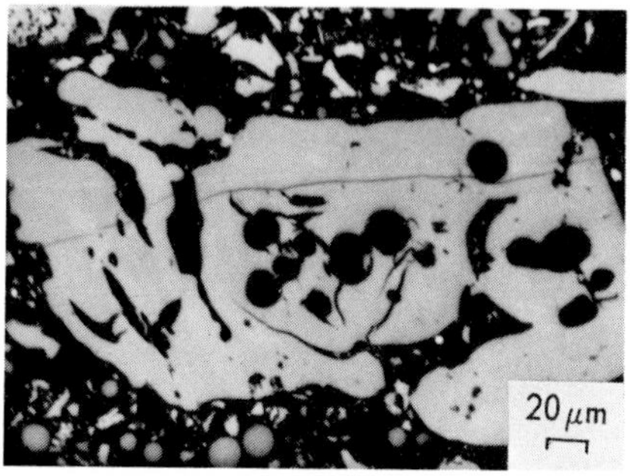

Fig. 10.   The apparent plasticity and partial reaction of semifusinite from the 400°C batch hydrogenation run using the Australian Callide coal.

The classic study by Fisher *et al.* (2) and later work by Davis *et al.* (3) indicate that fusinite is less suitable than the other "inertinites" for hydrogenation and may in fact be totally inert during liquefaction. It is observed in varying concentration in hydrogenation residues depending upon the fusinite concentration in the original feed coal. As a result of mechanical abrasion in the turbulent flow of continuous-flow reactors, fusinite is generally broken into minute particles. It is usually readily apparent in this dispersed form as high reflecting angular fragments which often exhibit remnant cell structure.

Semifusinite is transitional in reflectance and morphology between vitrinite and fusinite. Partial reactivity has been assumed for semifusinite during carbonization, and may

be valid for the liquefaction process. The subbituminous
Australian Callide coal which has an unusually high concentration of semifusinite (Table 1), was reacted in the batch
hydrogenation system to investigate semifusinite's reactivity.
The coal gave a 40 percent conversion at 400°C indicating the
probability that components other than vitrinite and exinite
were involved in the reaction. Optical examination of the
Callide residue confirmed the partial reaction of semifusinite. Fig. 10 shows semifusinite with slightly rounded edges
and irregularly shaped cell lumens which may have begun to
close when the structure became plastic. The residue also
contains a high concentration of spherical vitroplast. Since
this coal has a very low vitrinite content, and noting the
close association of the vitroplast with the plasticized
semifusinite, it seems likely that this vitroplast has been
derived from semifusinite. At this stage of research the
reactivity of semifusinite must be determined by empirical
testing of individual coals.

Another coal constituent which is considered to be inert
during carbonization is macrinite. Table 1 reports a high
percentage of this component in the Indiana #1 Block feed
coal. Batch hydrogenation of this sample at 450°C resulted
in a 72 percent conversion, indicating again that some of the
"inert" constituents must have reacted. Microscope observation of the progressive reaction residues suggests that with
increasing temperature, macrinite reacts and that by 425°C
no distinguishable residue analog remains. By including
macrinite with the reactive constituents (vitrinite and
exinite, Table 1) the percentage of reactive macerals now
closely corresponds to the 72 percent total conversion. As
with semifusinite, the relative reactivity of macrinite must

A constituent which is present in low concentration in
this series of high volatile bituminous coals is micrinite.
Owing to its fine grain size (0.1-1.0µm), the behavior of
micrinite during hydrogenation is difficult to assess. There
are physical similarities between micrinite and the granular
material produced during hydrogenation by the degradation of
sporinite and weakly reflecting vitrinite. Whether micrinite
is reactive or contributes to the insoluble residues of
hydrogenation must be determined by compositional differentiation of the granular residue category. That is, a relatively low concentration of organic species in the submicron
granular residue derived from the hydrogenation of a coal
containing a moderate amount of micrinite, would indicate that

this maceral has been reactive. At present, however, the
reactivity of micrinite is unknown.

E.   Carbonized Residue Constituents

The development of anisotropy in the insoluble organic
constituents of hydrogenation can be extensive. Any one of
the residue components described above can develop anisotropic domains and, at the same time, retain a characteristic
morphology. The cenosphere in Fig. 6 is an example. The
outer wall has been carbonized to a fine-grained anisotropic
mosaic. Similarly, vitroplast may become anisotropic depending on temperature and time. The distinction between this
type of semi-coke and that produced as a result of mesophase
development (nematic liquid crystal adsorption onto inert
surfaces) is important, although both types of semi-coke represent repolymerization and a loss of reactive constituents.
The appearance of these carbonized particles may indicate
that reaction conditions were too severe, or that reactor
flow was not efficient and that these particles were subjected to a longer residence time than reported.

IV.   SUMMARY AND CONCLUSIONS

As a result of optical studies of the residues obtained
from a series of hydrogenation experiments at progressively
higher temperatures, insight is gained into the origin of
residue components from continuous-flow liquefaction. Microscopic observations show the relative uniformity of residue
components between different feed coals and hydrogenation
processes. The result of these studies is a classification
of organic solid residues (Table 2), and some understanding
of the thermal alteration of coal macerals.
Diversity of thermal environments during liquefaction is
clearly reflected by these residues. The appearance of unreacted or slightly altered vitrinite implies that temperatures are too low. In contrast, the occurrence of anisotropic
semi-coke suggests more severe conditions. The slight variation in reflectance of the vitroplast spheres (Figs. 2-5) may
also indicate that temperatures in the reaction vessel are
variable. The lower reflecting spheres apparently have been
less thermally altered than those of higher reflectance.
The chemical heterogeneity of coal macerals is reflected
in the morphologies and phases observed in the insoluble

residues. The fact that some vitrinitic particles become plastic while others are broken down into very fine materials and disseminated during hydrogenation is an example of chemical heterogeneity. Microscopic examination of coal indicates that there are slight differences in reflectance and morphology between vitrinitic particles within the same sample. These differences have become important in predicting the reactivity of a coal during coke making and may be similarly applicable to predicting a coal's reactivity during hydrogenation. The partial reaction of semifusinite for some coals is a further demonstration of the effects of chemical heterogeneity.

The presence of certain residue components could affect adversely the efficient operation of continuous-flow processes. Besides the formation of mesophase on inert surfaces as a coking mechanism during hydrogenation, the fine grained granular residue and spherical vitroplast components represent a potential problem. Large concentrations of these phases could detrimentally affect filtration. Their fine grain size and nearly spherical shape could lead to a very tight packing during filtration, resulting in decreased filtration rate.

The full potential of residue microscopy probably has not yet been realized. However, by classification of the organic and inorganic insoluble constituents, a means of systems product balance may be developed. Comparison of quantitative point-count data derived from this classification should also aid in determining the optimum-yield run conditions.

V. ACKNOWLEDGMENTS

The research reported here is supported by the Electric Power Research Institute and the RANN Division of the National Science Foundation.

VI. REFERENCES

1. Guin, J.A., Tarrer, A.R., Taylor, Z.L., and Green, S.C., ACS Div. Fuel Chem. 20(1), 66 (April, 1975).
2. Fisher, C.H., Sprunk, G.C., Eisner, A., O'Donnel, H.J., Clarke, L., and Storch, H.H., U.S. Bur. Mines, Tech. Prog. Rep. 642, 1 (1942).

3. Davis, A., Spackman, W., and Given, P.H., Energy Sources 3, 55 (1976).
4. Teichmüller, M., Fortschr. Geol. Rheinl. Westfalen 24, 37 (1974).
5. Newall, H.E., and Sinnatt, F.S., Fuel 3, 424 (1924).
6. Street, P.J., Weight, R.P., and Lightman, P., Fuel 48, 342 (1969).
7. Neavel, R.C., Proc. Symp. Coal Agglom. Conversion Coal 119, (May, 1975).
8. Brooks, J.D., and Taylor, G.A., in "The Chemistry and Physics of Carbon" (P.L. Walker, Jr. Ed.), Vol. 4, p. 243. Marcel Dekker, New York, 1968.
9. Marsh, H., Dachille, F., Lley, M., Walker, P.L., and Whang, P.W., Fuel 52, 253 (1973).
10. Ammosov, I.L., Eremin, I.V., Sukhenko, S.I., and Oshurkova, L.S., Koks Khim. 12, 9 (1957).

INDEX

A

Analysis
  carbene-carboid, 236
  chromatographic, 247
  coals tested, 105
  COED oil, 235
    product, 225
  hydrogenation product, 40
  mass spectrometer, 221
  metals in oils, 238
  method of, 223
  monoaromatics, 225
  procedure, 234
  resins, 236
  saturates, 225
  short contact product, 159
  Synthoil product, 226, 235
Apparatus
  analytical, 222
  autoclave, 104, 122
  catalytic, flow, 65
  filtration, 187, 204
  H-coal, 4
  hydrogen solubility, 136
  liquefaction, 47
  liquid chromatographic, 246
  petrographic, 256
  settling, 174
  short contact time, 155
  two-step hydrogenation, 22
Aromatics, polycyclic, yields, 31, 37
Asphaltene
  conversion, 110
  formation, 108
  molecular weight, 111
  yields, 28

B

Bergius process, 1
Biphenyl, 64
n-Butylnaphthalene, 72
n-Butyltetralin, 72

C

Catalyst
  attrition test, 98
  carbon deposition, 93
  chemical stability, 82
  Co-Mo, 103
  coal mineral, 45, 246
  Co-Mo-Al, 246
  compounds, 80
  deactivation, 73, 92
  free energy of formation, 83
  metal deposition, 95
  oxides, 80
  screening, 47
  sintering by steam, 94
  thermal stability, 81
  titanium deposition, 96
Cenosphere development, 263
Cenospheres, 262
Chromatogram, phenanthrene
    products, 72
Chromatography
  analysis, 46, 137
  application, 223
  gel, application, 233
  liquid, application, 246
Coal
  ash composition, 96
  Black Mesa analysis, 3
  demineralized, hydrogenation, 52
  demineralized, hydrodesulfuri-
    zation, 54
  Illinois No. 6 analysis, 3
  intraparticle reactions, 169
  Kentucky, analysis, 105
  Kentucky No. 9/14, 48, 138
  Madisonville No, 9, analysis, 120
  mineral residue, 52
  optical analysis, 255
  petrography, 258
  solids agglomeration, 195
  solids coagulation, 175
  solution rate, 161
  West Kentucky, analysis, 156
  West Virginia, analysis, 105